NEUTRONS AND NUMERICAL METHODS—N_2M

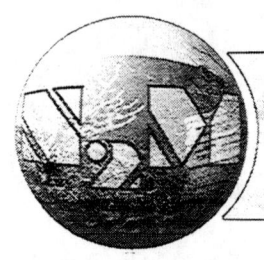

Neutrons and Numerical Methods

Editors: M. R. Johnson, G. J. Kearley, and H. G. Büttner

Institut Laue-Langevin, BP 156, F-38042 Grenoble Cedex 9

Sponsored by:

Institut Laue–Langevin
B.P. 156
38042 Grenoble, Cedex 9, France

&

Molecular Simulations Inc.
Parc Club Orsay Université, 20 rue Jean Rostand
91893 Orsay, Cedex, France

Acknowledgement

The organisers express their gratitude to Brigitte Aubert for her contribution as secretary of the workshop.

NEUTRONS AND NUMERICAL METHODS—N$_2$M

Grenoble, France   December 1998

EDITORS
M. R. Johnson
G. J. Kearley
H. G. Büttner
Institut Laue-Langevin, Grenoble

American Institute of Physics

AIP CONFERENCE PROCEEDINGS 479

Woodbury, New York

Editors:

M. R. Johnson, G. J. Kearley, and H. G. Büttner
Institut Laue-Langevin
BP 156
Cedex 9
F-38042 Grenoble
FRANCE

E-mail: johnson@ill.fr
kearley@ill.fr
buttner@ill.fr

The article on pp. 47–54 was authored by a U. S. Government employee and is not covered by the below mentioned copyright.

Authorization to photocopy items for internal or personal use, beyond the free copying permitted under the 1978 U.S. Copyright Law (see statement below), is granted by the American Institute of Physics for users registered with the Copyright Clearance Center (CCC) Transactional Reporting Service, provided that the base fee of $15.00 per copy is paid directly to CCC, 222 Rosewood Drive, Danvers, MA 01923. For those organizations that have been granted a photocopy license by CCC, a separate system of payment has been arranged. The fee code for users of the Transactional Reporting Service is: 1-56396-838-X/99/$15.00.

© 1999 American Institute of Physics

Individual readers of this volume and nonprofit libraries, acting for them, are permitted to make fair use of the material in it, such as copying an article for use in teaching or research. Permission is granted to quote from this volume in scientific work with the customary acknowledgment of the source. To reprint a figure, table, or other excerpt requires the consent of one of the original authors and notification to AIP. Republication or systematic or multiple reproduction of any material in this volume is permitted only under license from AIP. Address inquiries to Office of Rights and Permissions, 500 Sunnyside Boulevard, Woodbury, NY 11797-2999; phone: 516-576-2268; fax: 516-576-2499; e-mail: rights@aip.org.

L.C. Catalog Card No. 99-62727
ISBN 1-56396-838-X
ISSN 0094-243X
DOE CONF- 981213

Printed in the United States of America

CONTENTS

Preface viii
Group Photograph x

METHODOLOGY

Structure and Dynamics from First Principles 3
    M. C. Payne
Algorithms, Developments, and Applications in Molecular Modelling:
The GAMESS-UK *Ab Initio* Code 9
    M. F. Guest and H. J. J. van Dam
The Development of RMC Methods for Modelling Structural Disorder
in Crystalline Materials... 19
    R. L. McGreevy and A. Mellergård
Modelling the Thermal Expansion of Zeolites 28
    J. D. Gale
Multiple Scattering Effects in Deep Inelastic Neutron Scattering
Experiments.. 37
    J. Dawidowski, J. J. Blostein, and J. R. Granada
A Monte Carlo Tool for Simulations of Neutron Scattering Instruments ... 41
    L. L. Daemen, P. A. Seeger, T. G. Thelliez, and R. P. Hjelm
New Procedure for Multiple Scattering Correction 47
    M. Russina and F. Mezei

STRUCTURAL WORK

Structure of Multi-Component Glasses Using Diffraction Techniques
and Reverse Monte Carlo Modelling.................................... 57
    J. Swenson, L. Börjesson, and R. L. McGreevy
Microscopic Structure of Amorphous Carbon. Tight-Binding
Molecular Dynamics Study... 64
    S. Kugler, K. Koháry, and I. László
Monte-Carlo Sorption and Neutron Diffraction Study of the Filling
Isotherm in Clathrate Hydrates 70
    A. Klapproth, B. Chazallon, and W. F. Kuhs
Molecular-Dynamics Modelling and Neutron Diffraction Study
of the Site Disorder in Air Clathrate Hydrates 74
    B. Chazallon, A. Klapproth, and W. F. Kuhs
Neutron Scattering Studies of the Structure and Dynamics of Interlayer
Water and Hydrated Cations in Montmorillonite Clays 78
    D. H. Powell, M. Gay-Duchosal, and C. Pitteloud

*Italicized name indicates author who presented the paper.

Effect of Charge Transfer on the Local Order in Liquid Group IV Isoelectronic Compounds: Neutron Diffraction Data Versus Numerical Tight-Binding Simulations 83
    G. Prigent, J.-P. Gaspard, R. Bellissent, and C. Bichara

Geometric Frustration in Gadolinium Gallium Garnet: A Monte Carlo Study 90
    O. A. Petrenko and D. McK. Paul

Molecular Modelling of Organic Superconducting Salts 96
    S. A. French and C. R. A. Catlow

Potential Surface of Rotation-Translation Coupled Systems: $Me(NH_3)_6(PF_6)_2$, Me=Ni,Co 102
    P. Schiebel, H. G. Büttner, G. J. Kearley, M. Prager, and W. Prandl

Structure and Bonding in *Cis*-enol Systems 107
    G. K. H. Madsen

Simulations of Hydrogen Bonds in Crystals and Their Comparison with Neutron Diffraction Results 112
    S. J. Grabowski

LOW-FREQUENCY DYNAMICS

Calculation of Phonon Dispersion Curves by the Direct Method 121
    K. Parlinski

Phonons in Chalcopyrite Compounds 127
    P. Derollez, A. Laamyem, R. Fouret, B. Hennion, and J. Gonzalez

Frequency Dependent Specific Heat of Amorphous Silica: A Molecular Dynamics Computer Simulation 131
    P. Scheidler, W. Kob, J. Horbach, and K. Binder

The Boson Peak in Amorphous Silica: Results from Molecular Dynamics Computer Simulations 136
    J. Horbach, W. Kob, and K. Binder

Molecular Dynamics Simulation of Inelastic Neutron Scattering Spectra of Copper Azurin Hydration Water 142
    A. Paciaroni, A. R. Bizzarri, and S. Cannistraro

Analysis of Low-Frequency Motions in Proteins by Computer Simulation and Neutron Scattering 147
    G. R. Kneller and K. Hinsen

Simulation of Inelastic Neutron Scattering Spectra for Water Ice— A Most Effective Way of Testing Water Potentials 155
    J. Li and J. Tomkinson

Extracting the Vibrational Density of States from Neutron Scattering Data: Beyond the Effective Density of States 160
    S. R. Elliott, A. Haar, R. D. Oeffner, and S. N. Taraskin

*Italicized name indicates author who presented the paper.

INTERNAL VIBRATIONS AND TUNNELLING

**EXAFS Calculations Using Debye-Waller Factors Deduced
from Inelastic Neutron Scattering** 167
    M. L. *Hanham* and R. F. Pettifer
**Density Functional Theory and *Ab Initio* Methods Applied to the Analysis
of the Inelastic Neutron Scattering Spectra** 172
    A. Navarro, M. Fernández-Gómez, J. J. López-González, F. Partal,
    J. Tomkinson, and G. Kearley
**Search for a Reliable Nucleic Acid Force Field Using Neutron
Inelastic Scattering and Quantum Mechanical Calculations:
Bases, Nucleosides, and Nucleotides.** 179
    N. Leulliot, H. Jobic, and *M. Ghomi*
**NIS, IR, and Raman Spectra with Quantum Mechanical Calculations
for Analysing the Force Field of Hypericin Model Compounds** 183
    J. Ulicny, N. Leulliot, L. Grajcar, M.-H. Baron, H. Jobic, and *M. Ghomi*
**Low Frequency Internal Vibrations of Norbornane and its Derivatives
Studied by IINS and Quantum Chemistry Calculations.** 187
    K. Holderna-Natkaniec, I. Natkaniec, and V. D. Khavryutchenko
**Neutron Spectrometry and Numerical Simulations of Low-Frequency
Internal Vibrations in Solid Xylenes** 191
    I. Natkaniec, K. Holderna-Natkaniec, J. Kalus, and V. D. Khavryutchenko
**Molecular Dynamics Simulation of Inelastic Neutron Scattering
Spectra of Librational Modes of Water Molecules in a Layered
Aluminophosphate** 195
    A. J. Ramirez-Cuesta, P. C. H. Mitchell, S. F. Parker, A. P. Wilkinson,
    and P. M. Rodger
**On the Origin of the Distribution of Potential Barriers for
Methyl Group Dynamics in Glassy Polymers: Neutron Scattering
and MD-Simulations** 201
    F. Alvarez, A. Alegría, *J. Colmenero*, T. M. Nicholson, and G. R. Davies
**Density Functional Theory for the Calculations of the Rotational
Potentials of Methyl Groups.** 206
    B. Nicolai and G. J. Kearley
Rotation-Precession and Rotor-Rotor Coupling in 4-Methyl-Pyridine 212
    M. A. Neumann, M. Plazanet, and M. R. Johnson
**Tribromomesitylene Structure at 14 K: Methyl Conformation
and Tunnelling** 217
    F. Boudjada, J. Meinnel, A. Cousson, W. Paulus, M. Mani,
    and M. Sanquer

Conference Programme 223
List of Participants 225
Author Index 237

*Italicized name indicates author who presented the paper.

Preface

The workshop was attended by 90 scientists. While the majority of participants came from Europe, East and West, neutron centres in the United States (NIST and Los Alamos) were represented and the most distant visitors were from Japan and Argentina. But the most striking aspect of the participants as a whole was the average age, many of the senior scientists commenting on how many young post-grads and post-docs had come to the meeting.

The workshop began with a welcome by ILL's director, Dirk Dubbers, and then a talk from Mike Payne (Cambridge) on density functional theory, the subject of the 1998 Nobel Prize in Chemistry. While a large range of experimental areas was covered, most sessions had a common theme. The computational techniques covered the whole spectrum from Monte Carlo and molecular dynamics to *ab-initio* methods. Two poster sessions were well attended and a prize was awarded for the best poster, judging criteria being scientific content, presentation and enthusiasm of the author. At the end of the conference dinner, held in the historic surroundings of the 17th century Château de Sassenage, Peter Scheidler (Mainz) was awarded a Santa Claus filled with chocolate eggs and a bottle of Chartreuse for his poster on the time and frequency dependence of heat capacity in silica.

Rob McGreevy (Uppsala) concluded the meeting after the Saturday morning session. Most of the presentations included a combination of neutron scattering and numerical modelling, several of the invited computational chemists commented on how well this complementarity of techniques has already been exploited. In other cases only experimental work was presented, a number of participants had come to obtain an insight into the range of modelling techniques available and their applicability to different research areas. While the ever-increasing power of computers and the availability of software will ensure that numerical modelling can only become more widespread, neutron scattering has a fundamental role to play in providing a direct experimental benchmark to test simulations. Finally the take-home message from the workshop was that *the marriage of neutrons and numerical methods enables us to make the essential step from data analysis to data understanding.*

The proceedings have been divided into four chapters, with the usual problem of some articles falling between the classifications. In the first chapter, the articles concern "methodology". Articles in which the scientific objective relates to an aspect of structure determination are grouped together in the "structure" chapter. The last two chapters concern dynamical studies. These are separated into "low-frequency dynamics" which are typically collective in character and "internal vibrations and tunnelling" which are local excitations.

Editors: M. R. Johnson, G. J. Kearley, and H. G. Büttner

Grenoble, March 1999

Participants of the N$_2$M Workshop at the main entrance of the ILL.

METHODOLOGY

Structure and Dynamics from First Principles

M.C. Payne

*Cavendish Laboratory, University of Cambridge,
Madingley Road, Cambridge, CB3 0HE, UK.*

Abstract. The combination of developments in theoretical numerical and computational techniques has made it possible to study a wide range of scientific problems from first principles. In this paper I shall review the capabilities of first principles calculations, highlighting both their strengths and weaknesses.

1. INTRODUCTION

First principles or ab initio calculations allow one to compute the energy of a system of atoms by solving the quantum mechanical equations for the electrons in a system. Thus, at least in principle, any physical or chemical property of a system that is related to a total energy or, more usually, to a difference between total energies can be computed taking as input only the atomic numbers of the atoms in the system. However, it is impossible to compute the exact many-electron wavefunction for a system containing more than a few electrons and so direct solution of the Schrödinger equation does not provide a technique for studying complex systems. Density functional theory reformulates the interacting electron problem and provides a tractable technique for performing first principles calculations for complex systems. Density functional theory will be reviewed in the following section where a description of the approximations necessary for its practical implementation will be given. The consequences of these approximations for the problems which are suitable for investigation using density functional theory calculations will also be discussed. In section 3, I shall describe the system sizes and timescales that are presently accessible to density functional theory calculations and briefly discuss how these might increase in the future. In section 4, I shall briefly describe applications of density functional theory calculations to investigate the reaction of methanol in zeolite catalysts and to study the aluminium (110) surface at finite temperature.

2. DENSITY FUNCTIONAL THEORY

The foundation of density functional theory (DFT) is the Hohenberg-Kohn theorem (1) which proves that no two groundstate wavefunctions, ψ_1 and ψ_2, of different Hamiltonians, H_1 and H_2, can generate the same single particle density $\rho(\mathbf{r})$. This theorem proves the existence of a unique functional of the single particle electron density, $F[\rho(\mathbf{r})]$ such that for a system specified by an external potential acting on the electrons $U(\mathbf{r})$ an energy functional $E[\rho(\mathbf{r})]$ can be written

$$E[\rho(\mathbf{r})] \;=\; \int U(\mathbf{r})\rho(\mathbf{r})d^3\mathbf{r} \;+\; F[\rho(\mathbf{r})] \tag{1}$$

such that the minimum value of $E[\rho(\mathbf{r})]$ is the exact groundstate energy of the system and the density that minimises the functional, $\rho_0(\mathbf{r})$, is the exact groundstate density. Thus, in principle, density functional theory allows us to calculate exact groundstate energies and electronic densities for any system.

To make progress with density functional theory we need to determine the functional $F[\rho(\mathbf{r})]$. This contains all the information about the quantum mechanics of the electron system and so we can expect it to be extremely complicated. Recent work has shown that it is possible to approximate this entire functional in real materials provided that the electron density in the material does not vary too rapidly (2). However, most density functional theory calculations use a different approximation to $F[\rho(\mathbf{r})]$, based on the work of Kohn and Sham (3). They divided $F[\rho(\mathbf{r})]$ into three terms: the kinetic energy of the density $\rho(\mathbf{r})$; the Coulomb energy of the charge density interacting with itself, historically this is usually referred to as the Hartree energy; and the exchange and correlation energy associated with the density $\rho(\mathbf{r})$. Although it is correlation that introduces the complexity into the many-electron wavefunction, it is the kinetic energy that is the hardest to approximate. Kohn and Sham circumvented this problem by introducing a set of auxiliary, single electron-like orbitals ψ_n, that generate the charge density $\rho(\mathbf{r})$ so that the kinetic energy can be computed exactly from these wavefunctions. Now that the kinetic energy term has been dealt with the only thing left to do is to find a good approximation for the exchange-correlation functional $E_{xc}[\rho(\mathbf{r})]$. The simplest form for this functional is a purely local approximation in which it is assumed that the exchange-correlation energy density at a particular point in a system is the same as in a uniform electron gas whose density is the same as the electron density at that point. Hence, in this approximation, which is usually referred to as the local density approximation or the LDA,

$$E_{xc}[\rho(\mathbf{r})] \;=\; \int \varepsilon_{xc}^{ueg}(\rho(\mathbf{r'}))\rho(\mathbf{r'})d^3\mathbf{r'} \tag{2}$$

where $\varepsilon_{xc}^{ueg}(\rho(\mathbf{r'}))$ is the exchange-correlation energy density of a uniform gas of density $\rho(\mathbf{r'})$. These energies have been computed by Ceperley and Alder using quantum Monte Carlo calculations (4).

The local density approximation gives a remarkably accurate description of many materials properties such as equilibrium geometries and elastic moduli, which are typically predicted to an accuracy of a few percent. Energies are not so well predicted in the local density approximation and new functionals which include terms involving the gradient of the electron density have been developed that provide a more accurate description of the energies of most systems. These functionals are usually referred to as generalised gradient approximations or GGAs. However, it should be remembered that the LDA and GGA are both approximations to the true density functional and the predictions of physical properties made using these approximations are not exact. In some cases the errors are much larger than a few percent. Thus it is critical to only apply DFT calculations to problems where this error can be tolerated. Resolving very small energy differences between radically different systems, where systematic errors of DFT cannot be expected to cancel, is unlikely to be successful. Unfortunately, the errors associated with any particular functional cannot be quantified and so it is impossible to specify in advance whether a DFT calculation is appropriate or not for any particular problem.

One of the drawbacks of the Kohn-Sham approach is that it replaces a search over single particle densities to find the minimum energy, for which the computational cost would increase linearly with the number of atoms in the system, by a search over the auxiliary orbitals. As the auxiliary orbitals have to orthogonalised the computational cost scales as the cube of the number of atoms in the system. This has severe implications if we wish to study very large systems. Finally, it should be remembered that DFT applies to the electronic groundstate and so any physical or chemical property associated with excited states is not accessible to DFT calculations, one obvious example of this is the inability to correctly calculate bandgaps using DFT.

3. ACCESSIBLE SYSTEM SIZE AND TIMESCALES

DFT provides a tractable scheme for performing first principles calculations but the requirement of large numbers of basis functions per atom for accurate calculations and the cubic scaling of the computational cost still restrict the largest DFT calculations to systems containing hundreds of atoms. At one time only calculations using plane wave basis sets and pseudopotentials to describe the ionic potentials running on large parallel computers could be applied to systems this large. More recently, quantum chemistry approaches using localised basis states are now approaching the same capability. With careful choice of system a wide range of scientific problems can be addressed using systems of this size, particularly if the long range interactions are separated from the short range interactions. Long range interactions are essentially classical and do not require detailed quantum mechanical calculation. A good example is the separation of the energy of a dislocation into a core energy and a long-range elastic energy. The former can be calculated using first principles calculations on modest sized systems and the long-range terms can be dealt with using classical elasticity theory.

In the case of timescales accessible to first principles calulations, the situation is not so satisfactory. At most, timescales of the order of picoseconds are accessible. It might be thought that since the computational time scales linearly with the length of the simulation, much longer timescales will become accessible as more powerful computers become available. Unfortunately, this may not be the case since most first principles dynamical simulations suffer from a loss of energy from the ionic system. While this loss is modest over a simulation time of a picosecond it is not clear that first principles dynamical simulations could be extended to the timescale of nanoseconds.

First principles calculations are also extremely limited in the number of configurations that can be explored. While they can very effectively distinguish the relative energies of a small number of different possible atomic configurations they are not powerful enough to perform an exhaustive search for low energy configurations in a complex configuration space. Thus, although in principle ab initio calculations take as input only the atomic numbers of the constituent atoms, in reality a reasonable model of the connectivity of the atomic structure is also required. However, the exact position of every atom of the system is not required since the calculations can very effectively locate the nearest local energy minimum for each starting structure. This is extremely important since no experimental technique can determine exact atomic positions in complex structures and so it is important that the ab initio calculations can perform this task.

It is possible that DFT calculations might be performed for significantly larger systems in the future if linear scaling techniques can be implemented (5). These techniques are already widely used in tight binding approaches but have proved much more difficult to implement in first principles schemes. It should be remembered that this techniques will do nothing to overcome the problems of limited timescales or poor exploration of complex configuration spaces, indeed these problems will be even more severe since the relevant timescales and complexity of the configuration space increases rapidly with increasing system size.

4. APPLICATIONS

Despite the cautionary note of the last two sections, it is important to appreciate just how large an impact DFT calculations are making across a wide range of scientific disciplines. There is an enormous range of scientific problems that are accessible to first principles investigation provided one is careful about the choice of problem so that the required system size and timescale are not too large and the errors inherent within practical implementations of DFT do not invalidate the results of the calculations. I shall describe two areas where first principles calculations have helped to progress scientific understanding.

4.i. Reaction of Methanol in Zeolite Catalysts

Figure 1 shows a single methanol molecule in a unit cell of the zeolite ferrierite which contains a single Brønsted acid site. In this geometry, the proton spontaneously transfers from the zeolite framework to the methanol molecule to create a chemisorbed methoxonium ion. At one time it was believed that this proton transfer was responsible for the catalytic properties of the zeolites. However, our calculations have shown (6) that this proton transfer is not associated with a weakening of the bond between the carbon and oxygen atoms in the methanol, and so this process cannot account for the catalytic action of zeolites. Further simulations performed with larger numbers of methanol molecules per acid site have shown that at these higher loadings there can be significant elongation of the C-O bond. A transition state search has shown that the activation energy for breaking the C-O bond is indeed reduced (7). This is an example where first principles calculations have significantly advanced the understanding of a scientific problem of such complexity that experiment alone had begun to reach the limit of its usefulness.

4.ii. Finite Temperature Properties of the Al(110) Surface

Figure 2 shows the layer averaged mean squared displacements for the first four layers of the Al(110) surface at 400K from a first principles calculation on a 72 atom unit cell, consisting of an 8 layer slab with 9 atoms per layer. The time in this figure indicates the point in the simulation at which the thermal average was started, thus later times represent averages over shorter runs. It can be seen that only one or two picoseconds are required to extract these mean squared displacements and it can be

FIGURE 1. Methanol in the zeolite ferrierite, containing a single Brønsted acid site.

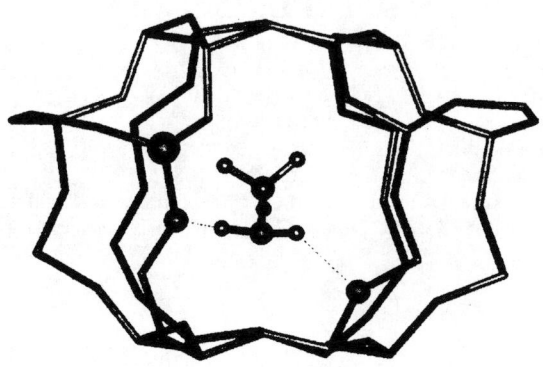

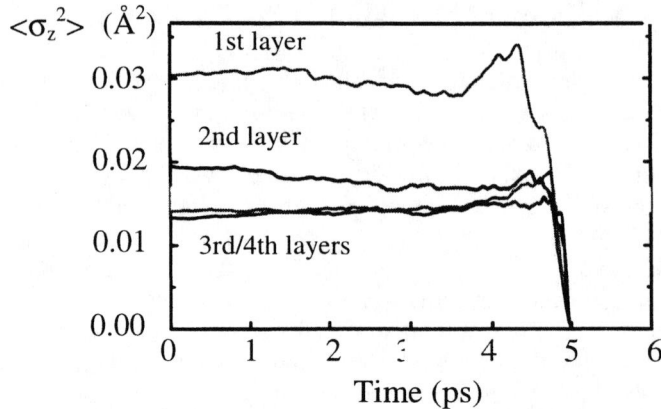

FIGURE 2. Layer averaged mean squared displacements normal to the Al(110) surface at 400K as a function of starting time in simulation. The shaded area indicates the experimental value for bulk Al.

seen that values for the lower layers agree with experimental measurements for bulk Al. The remarkable result that can be seen in figure 2 is that the second layer atoms have a larger vibrational amplitude normal to the surface than the surface atoms. This finding can be rationalised from an understanding of metallic bonding and the fact that the second layer atoms lie between the first layer atoms. This combination of factors produces a very shallow potential energy surface for motion of the second layer atoms upwards between the first layer atoms. The large anharmonic motion of the second layer atoms that results from this combination of factors actually produces a reduction in the spacing between the first and second layers with increasing temperature.

REFERENCES

(1) Hohenberg, P., and Kohn, W., *Phys.Rev.* **136**, B864 (1964).
(2) Smargiassi, E., and Madden, P.A., *Phys.Rev.B* **49**, 5220 (1994).
(3) Kohn, W., and Sham, L.J., *Phys.Rev.* **140**, A1133 (1965).
(4) Ceperley, D.M.., and Alder, B.J., *Phys.Rev.Lett.* **45**, 566 (1980).
(5) Li, X.P., Nunes, W., and Vanderbilt, D., *Phys.Rev.B* **47**, 10891 (1993).
(6) Stich, I., Gale, J.D., Terakura, K., and Payne, M.C., *Chem.Phys.Lett.* **283**, 402 (1998).
(7) Gale, J.D., Sandré, E., and Payne, M.C., to be published in *Chem.Comm.* (1998).

Algorithms, Developments and Applications in Molecular Modelling: The GAMESS-UK *Ab Initio* Code

M.F. Guest and H.J.J. van Dam

Department for Computation and Information,
CCLRC Daresbury Laboratory, Daresbury,
Warrington WA4 4AD, Cheshire, UK

Abstract. Focusing on molecular electronic structure, an outline is presented of the range of methods commonly used in computational chemistry, with consideration given to the accuracy, limitations and performance costs of the HF, DFT, MP2 and CCSD methods in understanding neutron-scattering related phenomena. Using GAMESS-UK as an example of a large electronic structure package, we illustrate the capabilities of these methods in computing the geometrical structures of transition metal complexes. These studies both reinforce the value of DFT, and reveal an overall inconsistency in the MP2-based predictions. Finally, emphasis is given to the cost effectiveness and applicability of the spectrum of hardware available to the computational chemist, from the single-user PC to massively parallel processors (MPP). We illustrate the potential of the latter platforms in enhancing the level of desk-top simulations by two-orders of magnitude, and demonstrate the potential of parallel PC-based Beowulf Systems.

INTRODUCTION

Computational chemistry covers a wide spectrum of activities ranging from quantum mechanical calculations of the electronic structure of molecules, to classical mechanical simulations of the dynamical properties of many-atom systems, to the mapping of both structure-activity relationships and reaction synthesis steps. In many cases predicting physical observables is bounded by the available computer capacity. In this paper we give an outline of the capabilities of methods commonly used in studying molecular systems, i.e. self-consistent field (SCF)[1] Hartree Fock (HF), Density Functional Theory (DFT), Moller-Plesset perturbation theory (MP2) and Coupled Cluster theory (CCSD). In section 2 we consider these methods in relation to neutron-scattering phenomena, and consider limitations and performance using GAMESS-UK [1] as a typical example of a large electronic structure

---

[1] All acronyms are listed in the appendix.

TABLE 1. Comparison of Experiment and Theory; Bond Length Errors (Å) with a DZP Basis Set

Molecule	Coordinate	HF	MP2	CCSD	MP4	CCSD(T)
NH_3	N-H	-0.011	0.002	0.003	0.004	0.004
CH_4	C-H	-0.001	0.003	0.005	0.006	0.006
C_2H_2	C-H	-0.002	0.006	0.007	0.008	0.008
	C-C	-0.012	0.024	0.016	0.025	0.023
CH_2	C-H	-0.005	0.003	0.004	0.008	0.006
	C=O	-0.015	0.022	0.015	0.023	0.021
HCN	C-H	-0.003	0.004	0.005	0.006	0.006
	C-N	-0.017	0.032	0.019	0.030	0.024
Average Abs. Error(Å)		-0.009	0.013	0.009	0.015	0.013

package. An illustration of these capabilities in determining the structures of transition metal complexes is given in section 3. Finally, emphasis is given to the cost effectiveness and applicability of available hardware, from the single-user PC to massively parallel processors (MPP).

COMPUTATIONAL CHEMISTRY METHODS

A survey of abstracts from this conference shows the wide variety of phenomena studied by neutron scattering, including molecular structure, vibrational and rotational spectra, quantum tunneling, molecular dynamics, and magnetisation. The computational properties needed in the associated simulations include the molecular energy and energy gradients, the energy second derivatives, and spin densities. These can in general be obtained from either force field or *ab initio* calculations.

The computational resources needed to obtain these properties may be illustrated by the way the cost of an energy calculation scales with the system size. With the configuration interaction (CI), HF, DFT, and molecular dynamics methods this cost formally scales as N^6, N^4, N^3 and N^2 respectively (N is the number of atoms). In practice neglecting interactions between distant atoms reduces these scalings significantly, e.g. to $N^{2.25}$ for HF. Still the costs limit the size of tractable systems. Given that accuracy is known to be limited in all QC methods with very small basis sets, we have used polarized basis sets throughout this paper (at least DZP or 6-31G*). In Table 1 a comparison of experimental bond lengths with those predicted by theory is shown [2]. HF is the simplest and most widely used *ab initio* method and a wealth of information on its applicability as function of property and compound class is available [3,4]. Despite its simplicity HF bond lengths and angles are typically very accurate (to within 1%) [3,5]. The correlated methods show little improvement in structural predictions over HF (except when the single determinant approximation fails).

In Table 2 a comparison of the mean absolute errors in bond length, angles

TABLE 2. Mean Absolute Errors using a 6-31G* Basis Set

Property	SCF	MP2	QCISD	DFT B-LYP (9700 QC points)
Bond Length (Å)	0.02	0.014	0.013	0.020
Bond Angle	1.99	1.78	1.79	2.33
Frequency (cm^{-1})	168	99	42	73
Atomisation Energy (k/cal)	85.9	22.4	22.8	5.6

and harmonic frequencies for 32 molecules is shown [6]. HF vibrational frequencies for covalent bonds are too high by 10-12% (very low frequency modes have higher errors), while zero point vibrational energies are accurate to 1 kcal/mole if scaled by 0.9. The accuracy of energy differences is very much a function of the system or reaction under study. Conformational energies are usually accurate to 0-2 kcal/mole, while isodesmic reaction energies are accurate to 2-4 kcal/mole. Increasingly large errors are found when moving from protonation / deprotonation reactions (10 kcal/mole), to atomization and homolytic bond-breaking reactions (25-40 kcal/mole); HF computed reaction barriers also show large errors.

Although DFT [7,8] is in practice slightly more costly than HF, it has been found to be as accurate as MP2 or CCSD(T) for many properties including structures, reaction energies, and vibrational frequencies. While DFT corrects many qualitative failures of HF, it remains a method with potential shortcomings, e.g. (i) systematic corrections are not possible, (ii) the often poor treatment of transition states, (iii) the treatment of excited states, and (iv) the requirement for basis set, functional and integration grid.

The GAMESS-UK Electronic Structure Package

GAMESS-UK [1] represents a typical large electronic structure code, offering a variety of methods for treating molecules; other such codes include GAUSSIAN [9], CADPAC [10], GAMESS-US [11], MOLPRO [12], Turbomole [13], HONDO [14] etc. Present functionality of GAMESS-UK lies at varying levels of approximation, including SCF (RHF, UHF, CASSCF, MCSCF, GVB) and DFT, MP2, MP3, MP4, CCSD(T), CI (MRDCI, Direct-CI, Full-CI) and analytic SCF and MP2 second derivatives. The package is available on vector hardware (e.g. Cray Y-MP, C90 and J90), on workstations from all leading vendors (e.g. SGI, HP, IBM, Compaq/Digital, and SUN), on parallel hardware (e.g., Cray T3E, SGI Origin series, and the IBM SP series), and on PCs under Linux, FreeBSD and WindowsNT.

STRUCTURES OF TRANSITION METAL COMPLEXES

Accurate structural predictions of transition metal complexes from computation has been an area of long standing interest [15]. Much of the growing acceptance

TABLE 3. RMS Deviation of Calculated Metal-Ligand Bond Lengths from Experiment (Å)

Transition Metal Ligand	M-L	SCF	MP2	DFT S-VWN	DFT B-LYP	DFT B-P86	DFT B3LYP
Oxide	M-O	0.075	0.061	0.022	0.011	0.008	0.022
Fluoride	M-F	0.036	0.038	0.037	0.028	0.025	0.024
Chloride	M-Cl	0.069	0.037	0.026	0.045	0.023	0.027
Carbonyl	M-C	0.190	0.092	0.048	0.034	0.022	0.027
Hydride	M-H	0.095	0.127	0.046	0.036	0.039	0.043
Organometallic	M-C	0.136	0.101	0.051	0.047	0.014	0.029

by chemists of DFT is due to its success in treating these complexes. In contrast, HF geometry optimisations on transition metal compounds often lead to bond lengths that are longer than experiment [15]- [24], e.g. the Fe-C bond in ferrocene is 0.23 Å too long [16,17]. The role of electron correlation in such studies has been demonstrated by Siegbahn et al. [18]. In contrast to HF, CCSD and CCSD(T) have been shown to be very accurate, with CCSD(T) giving optimal structures for $Cr(CO)_6$ and $Cr(CO)_5$ [19]. Fewer structures from the less computationally demanding MP2 method have appeared; ECP-based studies find the Cr-CO bond in $Cr(CO)_6$ to be too short (1.883 vs 1.918 Å) [20]. An excellent demonstration of the success of DFT is given by Sosa et al [21], who applied LDF (using DMOL and DGauss) to 45 complexes. The results reveal a RMS deviation in bond lengths of 0.026 Å with experiment; using gradient corrected functionals, DFT matches CCSD(T) in quality [22].

We have extended these previous findings through a study [24] of the equilibrium geometries of 60 transition metal complexes (containing oxygens, halogens, alkyl and aryl groups, carbonyls, nitrosyls etc.) employing HF, MP2 and DFT using DZ, DZP and cc-PVDZ basis sets. All SCF, MP2 and DFT calculations with the S-VWN, B-LYP, and B3LYP functionals were performed using GAMESS-UK; DFT calculations using the B-P86 functional employed NWChem [see section 4]. All calculations were run on the parallel IBM-SP systems at Daresbury and the Pacific Northwest National Laboratory (PNNL). Summarizing the results, Table3 shows the RMS deviation between calculated and experimental bond lengths. The most interesting results are seen for the carbonyl, hydride and organometallic complexes; HF exhibits unacceptable errors, with the metal-C distance significantly overestimated in all carbonyl and organometallic complexes. While not apparent from the RMS values, MP2 leads to bond lengths *shorter* than experiment, e.g. by 0.17 Å for the hydrides. In contrast, DFT provides a far more systematic tool in structural predictions e.g., it consistently overestimates experiment by 0.03-0.05 Å when using the B-LYP functional. In agreement with previous results [23] we find that distances improve using the hybrid HF/DFT scheme due to Becke, and correlation functionals developed by Perdew in 1986 and 1991.

HARDWARE CONSIDERATIONS

Single Processor Performance; Workstations and PCs

While workstations have demonstrated their cost-effectiveness against vector supercomputers, recent developments with PCs suggest the position of the workstation as the desktop system of choice is now under threat. A benchmark featuring 12 GAMESS-UK applications [25] shows that the EV6-based DEC Alpha 8400/575 is the fastest machine, followed by IBM RS/6000-43P/M260 (67%), HP PA-9000/C240 (65%), SGI Origin2000/250 (55%), SUN HPC4500/336 (53%), and Pentium II/400 (39%), where the percentages show performance relative to the DEC Alpha. The PentiumII is found to be only 2.6 times slower than the fastest workstations, suggesting that PCs have become a highly cost-effective platform for the computational chemist.

MPPs; Grand Challenge and Throughput Applications

While the potential of massively parallel computers (MPPs) has long been recognised, it is now apparent that new, scalable, algorithms are needed to fully exploit this potential [26–28]. The High Performance Computational Chemistry group within PNNL's Environmental Molecular Sciences Laboratory [29] is developing a new generation of molecular modelling software to take full advantage of MPPs. These efforts have resulted in the package NWChem that includes a range of electronic structure and molecular dynamics functionality [27]. A major part of the success of NWChem stems from the development of the Global Array (GA) tools [30]; this toolkit provides an efficient and portable "shared-memory" programming interface for distributed-memory computers. With these tools codes can be developed faster and are more extensible and maintainable.

In contrast to NWChem, both SCF and DFT modules in GAMESS-UK are parallelised in a replicated data fashion (each node having a copy of all data structures), this limiting the size of a calculation to about 2,000 basis functions on 256 MByte nodes. The main parallelism in the SCF is in the computation of the integrals and construction of the Fock-matrix; diagonalisation of this matrix is based on the parallel PEigS module from NWChem [31]. Figure 1 shows the speedups achieved using GAMESS-UK on up to 128 Cray T3E/1200 processors for both SCF and DFT calculations. The SCF calculations on morphine used a 6-31G** basis of 410 functions, those on cyclosporin a 3-21G basis of 1000 functions, while the DFT calculations on cyclosporin used a 6-31G basis and the B3LYP functional. Speedups of 108, 95, and 106 are obtained on 128 nodes for the morphine SCF, the cyclosporin SCF, and cyclosporin DFT calculation, respectively.

Substantial modifications were needed to parallelise the MP2 gradients [32]. Specifically, the conventional integral transformation has been omitted, with the MO integrals generated by recomputing the AO integrals and stored in memory

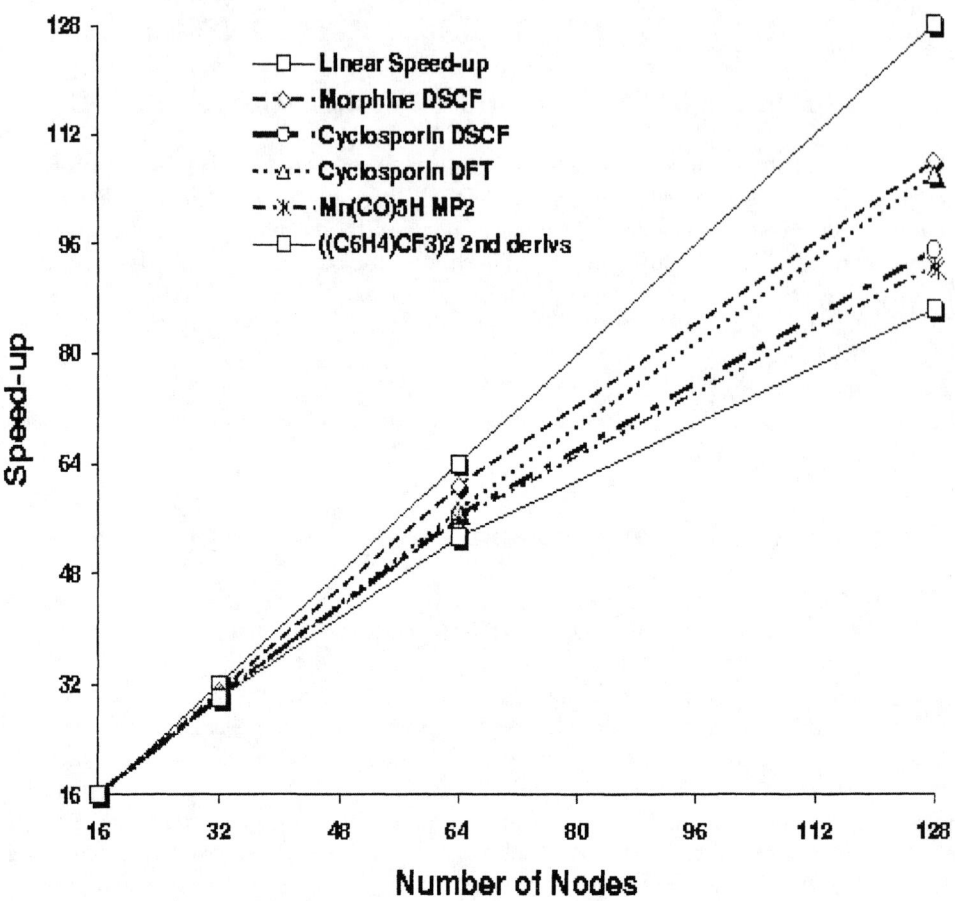

FIGURE 1. Parallel scaling of the GAMESS-UK direct-SCF, DFT, SCF second derivatives and MP2 gradient modules in calculations on Morphine, Cyclosporin, di(trifluoromethyl)-biphenyl and $Mn(CO)_5H$ on the Cray T3E/1200 (see text)

through the GA tools. Each node then computes the contributions resulting from its resident MO integrals. The figure shows a speedup of 93 achieved using 128 processors in a full MP2 geometry optimisation of the $Mn(CO)_5H$ molecule (217 basis functions and 5 energy and 5 gradient calculations). Finally, the second derivative module is parallelised in similar fashion to the MP2 gradients (i.e. using the GA tools to house the MO integrals), with the second derivative 2-electron integrals (the most costly serial step) parallelised in the same way as the 2-electron inte-

TABLE 4. Time in Wall Clock Seconds for Direct-SCF and DFT Calculations on the Morphine molecule using GAMESS–UK

Calculation	Nodes	Machine		
		Cray T3E/900	Cray T3E/1200	Beowulf System
Direct-SCF	4	3434	2718	4775
	8	1740	1380	2617
	10	1369	1079	2179
DFT B3LYP	4	5128	4236	7700
	8	2581	2129	4187
	10	2036	1678	3448

grals. In a calculation of the frequencies of 2,2'-di(trifluoromethyl)-biphenyl using a 6-31G basis of 196 functions, we find a speedup of 86 using 128 processors.

Exploring the Potential of Beowulf Class Systems

An emerging class of highly cost effective mid-range machines include the so-called Beowulf systems [33], modestly sized parallel machines constructed by networking PC's running the Linux operating system. Their main drawback is the poor performance of the conventional inter-process communication mechanisms; to investigate the potential of such systems we have recently installed a 10 PentiumII CPU Beowulf at Daresbury [34]. The present parallel version of GAMESS-UK on this system is MPI-based since the message latency and bandwidth are too poor for the GA-tools. In Table 4 we show the performance of the SCF and DFT modules in calculations on the morphine molecule (6-31G** basis with 410 functions) on the Cray T3E/900, Cray T3E/1200 and on the Daresbury Beowulf systems. The total times to solution show that the Cray T3E/1200 is only twice as fast as this Beowulf system on 10-nodes. This performance should be viewed in the light of, (i) the Beowulf scaling which is, as expected, much worse than on the T3E, and (ii) larger calculations will need the GA tools for parallel matrix algebra. Nevertheless these results are most encouraging; to be competitive with ten nodes of a T3E on a system costing less than a mid-range workstation points to the clear potential of these systems.

APPENDIX: GLOSSARY OF ACRONYMS

AO Atomic Orbital

B-LYP density functional constructed from the Becke exchange functional proposed in 1988 and the Lee, Yang, and Parr correlation functional

B-P86 density functional constructed from the Becke exchange functional proposed in 1988 and the correlation functional proposed by Perdew in 1986

B3LYP density functional constructed from Becke's 3 parameter exchange functional and the Lee, Yang, and Parr correlation functional

CASSCF Complete Active Space Self Consistent Field method

cc-PVDZ correlation consistent basis set including Polarisation functions describing the Valence orbitals with a Double Zeta set of functions

CCSD Coupled Cluster theory including all Single and Double excitations

CCSD(T) Coupled Cluster theory including all Single and Double excitations and accounting for Triple excitations through perturbation theory

CI Configuration Interaction

DFT Density Functional Theory

DZ Double Zeta basis set

DZP Double Zeta basis set with Polarisation functions

ECP Effective Core Potential approximation

GA Global Array toolbox

GVB Generalised Valence Bond theory

HF Hartree-Fock theory/approximation

LDF Local Density Functional

MCSCF Multi Configurational Self Consistent Field method

MO Molecular Orbital

MPn Møller-Plesset perturbation theory upto n-th order

MPP Massively Parallel Processor

MRDCI Multi-Reference Direct Configuration Interaction method

PC Personal Computer

PEigS Parallel Eigenvalue Solver

PNNL Pacific Northwest National Laboratory

QC Quantum Chemistry

QCISD Quadratic Configuration Interaction method including all Single and Double excitations

RHF Restricted Hartree-Fock approximation (the same orbitals for α and β spin electrons)

RMS Root Mean Square

S-VWN Spin polarised correlation functional as proposed by Vosko, Wilk and Nusair

SCF Self Consistent Field method

UHF Unrestricted Hartree-Fock approximation (different orbitals for α and β spin electrons)

REFERENCES

1. GAMESS–UK is an *ab initio* package written by M.F. Guest, J.H. van Lenthe, J. Kendrick, K. Schoffel, P. Sherwood and H.J.J. van Dam, with contributions from R.D. Amos, R.J. Buenker, M. Dupuis, N.C. Handy, I.H. Hillier, P.J. Knowles, V. Bonacic-Koutecky, W. von Niessen, R.J. Harrison, A.P. Rendell, V.R. Saunders, and A.J. Stone.
2. see Table 8 of R.J. Barlett and J.F. Stanton, in Reviews in Computational Chemistry Vol. 5, Ed. by K.B. Lipkowitz and D.B. Boyd. p.65, New York: VCH Publishers, Inc. (1994).
3. W.J. Hehre, L. Radom, P. v.R. Schleyer, and J.A. Pople, Ab Initio Molecular Orbital Theory, Wiley, New York (1987).
4. H.F. Schaefer, J.R. Thomas, Y. Yamaguchi, B.J. DeLeeuw, G. Vacek, in Modern Electronic structure theory, D.R. Yarkony (ed.), World Scientific (1995).
5. W.A. Lathan, L.A. Curtiss, W.J. Hehre, J.B. Lisle and J.A. Pople, Prog. Phys. Org. Chem. **11** (1974) 175.
6. N.C. Handy in: Lecture Notes in Chemistry, **64**, Ed. B.O. Roos, Springer Verlag (1994) 91.
7. P. Hohenberg and W. Kohn, Phys. Rev. B. **136** (1964) 864. W. Kohn and L.J. Sham, Phys. Rev. A. **140** (1965) 1133.
8. E. Wimmer, Density Functional Methods in Chemistry, J.K. Labanowski and J.W. Andzelm, eds. Springer-Verlag, Berlin (1991) 7.
9. M. Frisch et al., Gaussian, Inc., 4415 Fifth Avenue, Pittsburgh, USA (1992).
10. R.D. Amos et al., CADPAC, Issue 6, University of Cambridge, (1995).
11. M.W. Schmidt, et al., QCPE Bulletin **7** (1987) 115.
12. MOLPRO is an *ab initio* package written by H.-J. Werner and P.J. Knowles.
13. R. Ahlrichs, M. Br, M. Hser, H. Horn and C. Klmel, Chem. Phys. Letters (1989) 162.
14. M. Dupuis, J.D. Watts, H.O. Villar and G.J.B. Hurst, Comput. Phys. Commun. **52** (1989) 415.
15. M.F. Guest, in Lecture Notes in Chemistry, **44**, Ed. M. Dupuis, Springer Verlag (1986) 98.
16. H.P. Luthi, J.H. Ammeter, J. Almlof and K. Korsell, Chem. Phys. Letts. **69** (1980) 540; H.P. Luthi, J.H. Ammeter, J. Almlof and K. Faegri, J. Chem. Phys. **77** (1982) 2002.

17. J. Almlof, K. Faegri, B.E.R. Schilling and H.P. Luthi, Chem. Phys. Letts. **106** (1984) 266.
18. H.P. Luthi, P.E. Siegbahn and J. Almlof, J. Phys. Chem. **89** (1985) 2156.
19. L.A. Barnes, B. Liu and R. Lindh, J. Chem. Phys. **98** (1993) 3978.
20. A.W. Ehlers and G. Frenking, J. Amer. Chem. Soc. **116** (1994) 1514.
21. C.Sosa, J. Andzhelm, B.C. Elkin, E. Wimmer, K.D. Dobbs and D.A. Dixon, J. Phys. Chem. 1992, **96** 6630 .
22. B. Delley, M. Wrinn and H.P. Luthi, J. Chem. Phys. **100** (1994) 5785.
23. T.V. Russo, R.L. Martin and P.J. Hay, J. Chem. Phys **102** (1995) 8023.
24. M.F. Guest. P. Sherwood, G.D. Fletcher, E. Apra and H.A. Nichols, J.Chem. Phys., (1998) accepted for publication.
25. M.F. Guest, in Proceedings of the Machine Evaluation Workshop, CCLRC Daresbury Laboratory, (November 1998).
26. R.A. Kendall, R.J. Harrison, R.J. Littlefield and M.F. Guest, in: Reviews in Computational Chemistry, K.B. Lipkowitz and D.B. Boyd eds., VCH Publishers, Inc., New York (1994).
27. M.F. Guest, E. Apra, D.E. Bernholdt, H.A. Fruechtl, R.J. Harrison, R.A. Kendall, R.A. Kutteh, X. Long, J.B. Nicholas, J.A. Nichols, H.L. Taylor, A.T. Wong, G.I. Fann, R.J. Littlefield and J. Nieplocha, Future Generation Computer Systems **12** (1996) 273.
28. D.A. Dixon, T.H. Dunning, Jr., M. Dupuis, D. Feller, D. Gracio, R.J. Harrison, D.R. Jones, R.A. Kendall, J.A. Nichols, K. Schuchardt and T. Straatsma, in High Performance Computing, eds: R.J. Allan, M.F. Guest, A.D. Simpson D.S. Henty and D.A. Nichole, Proceedings of the HPCI'98 Conference, Plenum Publishing (1999).
29. the Environmental Molecular Sciences Laboratory, Pacific Northwest National Laboratory, Battelle Memorial Institute, see *http://www.pnl.gov:2080/*.
30. J. Nieplocha, R.J. Harrison and R.J. Littlefield, in: Supercomputing '94, IEEE Computer Society Press, Washington, D.C. (1994).
31. G. Fann and R.J. Littlefield, in: Sixth SIAM Conference on Parallel Processing for Scientific Computing (SIAM), (1993) 409.
32. G.D. Fletcher, A.P. Rendell and P. Sherwood, Molec. Phys. **91** (1997) 431.
33. D. Ridge, D. Becker, P. Merkey, T.Sterling and P. Merkey, Proceedings, IEEE Aerospace (1997).
34. the Daresbury Beowulf System see *http://www.dci.clrc.ac.uk/Activity/DISCO+926*.

The development of RMC methods for modelling structural disorder in crystalline materials

[1]R L McGreevy and [2]A Mellergård

[1]*Studsvik Neutron Research Laboratory, Uppsala University, S-611 82 Nykoping, Sweden.*
[2]*Materials Physics, Royal Institute of Technology, S-100 44 Stockholm Sweden.*

Abstract. We describe the development of reverse Monte Carlo methods for modelling structural disorder in crystalline materials from neutron powder diffraction data, explaining the advantages and disadvantages of the different methods available. The distinction between elastic (Bragg) scattering and total scattering measurements, and the different type of structural information that can be deduced from them, is made in some detail. The possibilities for further developments are described.

INTRODUCTION

Conventional crystal structure refinement/determination produces the time average structure of the material, and some averaged representation of deviations from it. Many materials, often of technological interest, have properties that depend on local correlations between atoms. This information must be accessed by some different technique, of which reverse Monte Carlo (RMC) modelling is one possibility.

RMC methods have been used for modelling structural disorder in crystalline materials for almost a decade [1]. The first applications were to fast ion conductors at high temperature, materials with a considerable degree of structural disorder, and hence methods already developed for the study of liquids and glasses [2,3] could be applied almost directly with only minor modifications. However some developments were needed to study the same materials at lower temperature, as the structures became more ordered [4]. It became clear from these studies that the amount of information that could be extracted from the resulting RMC models was limited because the effect of the finite experimental resolution was not considered. It also became clear that problems would be found for materials with magnetic scattering, even if this scattering was not of primary interest. For this reason we have developed a new RMC method, RMCPOW, which properly includes the experimental resolution and also allows the simultaneous modelling of both atomic and magnetic structures [5,6]. In this paper we describe the development of RMC methods for modelling structural disorder in crystalline materials, the advantages and disadvantages of the various methods available, and the requirements and prospects for further developments.

BRAGG AND DIFFUSE SCATTERING

In a neutron powder diffraction experiment one measures the number of neutrons scattered from a sample as a function of scattering angle relative to the incident beam. For simplicity we will restrict ourselves here to the case of a monochromatic incident beam though the principles are no different for time of flight techniques. At each point in the diffraction pattern the detector actually measures the total number of neutrons scattered as a function of angle (θ) relative to the incident beam. The intensity, $I(\theta)$, is therefore related to an energy integral of the dynamical structure factor, $S(\mathbf{Q},\omega)$, with the limits of the integral depending on the incident energy, E_0. The momentum transfer Q is a function of both θ and ω

$$I(\theta) \propto \left. \int_{-E_0}^{\infty} S(Q,\omega)\, d\omega \right|_{\theta} \tag{1}$$

Here we assume a monatomic sample for simplicity. $S(\mathbf{Q},\omega)$ is related to the van Hove correlation function $G(\mathbf{r},t)$, which contains information on the positions of atoms as a function of time.

$$G(\mathbf{r},t) = \frac{1}{N} \sum_{j,j'} \int <\delta(\mathbf{r}-\mathbf{r}'+\mathbf{R}_j)\, \delta(\mathbf{r}'-\mathbf{R}_{j'})> d\mathbf{r}' \tag{2}$$

$$S(\mathbf{Q},t) = \int e^{i\mathbf{Q}\cdot\mathbf{r}}\, G(\mathbf{r},t)\, d\mathbf{r} = \frac{1}{N} \sum <e^{-i\mathbf{Q}\cdot\mathbf{R}_j(0)}\, e^{i\mathbf{Q}\cdot\mathbf{R}_{j'}(t)}> \tag{3}$$

$$S(\mathbf{Q},\omega) = \int S(\mathbf{Q},t)\, e^{-i\omega t}\, dt \tag{4}$$

where the sample contains N atoms at positions $\mathbf{R}_j(t)$. $S(\mathbf{Q},t)$ is known as the intermediate scattering function. If the incident energy is much higher than any excitation energy in the sample then a constant angle integration is equivalent to a constant Q integration, and the lower limit of the integral in eq. (1) goes to infinity, so we obtain the total structure factor, $F(Q)$,

$$I(\theta) \propto F(\mathbf{Q}) = \int_{-\infty}^{\infty} S(\mathbf{Q},\omega)\,d\omega = S(\mathbf{Q},t=0) = \int e^{i\mathbf{Q}\cdot\mathbf{r}}\,G(\mathbf{r},t=0)\,d\mathbf{r}$$

$$= \frac{1}{N}\sum_{j,j'}<e^{-i\mathbf{Q}\cdot(\mathbf{R}_j(0)-\mathbf{R}_{j'}(0))}> = \frac{1}{N}\left|\sum_j e^{-i\mathbf{Q}\cdot\mathbf{R}_j(0)}\right| \quad (5)$$

For an isotropic sample (e.g. a powder) $F(Q)$ is related to the radial distribution function, $g(r) = G(\mathbf{r},t=0)$, by

$$F(Q) = 1 + \rho \int 4\pi r^2 \frac{\sin Qr}{Qr}(g(r)-1)\,dr \quad (6)$$

In conventional crystallography it is normally assumed that all sharp peaks in the scattering pattern represent elastic scattering, that is Bragg scattering. All other scattering in between or 'under' the sharp peaks may be either elastic or inelastic and will here be referred to as diffuse scattering - we make no distinction as to the source of that scattering. This diffuse scattering is simply subtracted and not used in the structural analysis. The elastic scattering only is then related to the elastic structure factor, $S(\mathbf{Q})$,

$$S(\mathbf{Q}) = S(\mathbf{Q},\omega=0) = \int S(\mathbf{Q},t)\,dt = \frac{1}{N}\sum_{j,j'}\int <e^{-i\mathbf{Q}\cdot\mathbf{R}_j(0)}e^{i\mathbf{Q}\cdot\mathbf{R}_{j'}(t)}>\,dt \quad (7)$$

To a first approximation the integral can be performed and the sum over the N atoms in the sample replaced by a sum over the N_c atoms in the unit cell, giving the more familiar expression

$$S(\mathbf{Q}) = \frac{1}{N_c}e^{-WQ^2}\left|\sum_k e^{-i\mathbf{Q}\cdot<\mathbf{R}_k(t)>}\right|^2 \quad (8)$$

where $<\mathbf{R}_k(t)>$ are now the time average positions of atoms in the unit cell and W is the Debye-Waller factor (DWF) which is related to the mean square deviation (msd) of atoms from their average positions.

We can therefore consider that the elastic structure factor gives us a picture of the time average structure of the material, and some measure of the average deviations from it, while the total structure factor gives us an instantaneous 'snap-shot' picture. Only at $T = 0$ are these two pictures the same. At other T the structure is not static but dynamic and so neither picture is 'correct'. Which picture is more useful depends on the problem being investigated. For most crystallographic studies the time average structure is the subject of interest and so elastic scattering is sufficient. In a liquid on the other hand all atoms are diffusing, so there is no time average structure and hence no strictly elastic scattering - only the instantaneous structure is meaningful. For crystalline structures with dynamical disorder, for example fast ion conductors, both time average and instantaneous structures may be of interest.

RMC MODELLING

The first RMC modelling of a crystalline material was done on the basis of the radial distribution function, obtained by direct Fourier transform of the total structure factor

$$g(r) = 1 + \frac{1}{(2\pi)^3 \rho} \int_0^\infty 4\pi Q^2 \frac{\sin Qr}{Qr} (F(Q) - 1) \, dQ \qquad (9)$$

This requires that the incident neutron energy be reasonably high so that

a) energy integration at a constant angle becomes a reasonable approximation to integration at constant Q (a small correction can be made for the deviation)

b) the lower energy limit of the integration is sufficiently large compared to any excitations in the sample and

c) the maximum momentum transfer is sufficiently large that oscillations in $F(Q)$ have died out, thus avoiding truncation of the transform in eq 9.

In practice (a) and (b) are reasonably well achieved with neutrons of wavelength 1 Å or lower. (c) is achieved for some highly disordered materials, this being 'helped' by the fact that the experimental resolution for short incident wavelengths tends to be rather poor at high scattering angles. However this immediately indicates that the information that can be extracted must be limited in some way by the resolution.

An alternative is to used a pulsed source diffractometer with a very short minimum wavelength and a correspondingly high maximum Q. However in this case the resolution tends to be rather good at backscattering angles and truncation can still be a problem, usually overcome by mutliplying the data by some form of modification (smoothing) function. The resolution then tends to be rather poor at low Q, and there are other detailed problems with the variation of resolution which there is not space to discuss here.

The problem of data truncation can be overcome if we model directly in Q space. This is also advantageous because one has a much better idea of likely errors (statistical and systematic) and their distribution in Q space rather than r space. However for a crystalline system, with long range order, the corresponding transform from $g(r)$ to $F(Q)$ is then truncated. This can be overcome by convoluting the experimental $F(Q)$ with the Fourier transform of a step function whose limit is the maximum r for which $g(r)$ can be calculated from the RMC configuration (i.e. half the minimum RMC box dimension) [4]. The effect of this convolution is illustrated in figure 1. In fact although it makes the data 'look' considerably worse we have found in tests that the results are the same as if the equivalent $g(r)$ were modelled, but only if the data contain no systematic errors. However if they do contain such errors then they are also convoluted and hence redistributed in $F(Q)$. A typical error such as a sloping background, which could not correspond to any physical structure and hence would tend to be 'ignored' by the RMC fit, will then become an oscillatory function that could correspond to some physical structure and hence will distort the RMC model. That being said, this Q space modelling does seem to be advantageous when the data contain some other sorts of errors, for example slight preferred orientation. This causes certain Bragg peaks to be more intense than they should be for a perfect powder, and gives an additional undamped oscillation in $g(r)$.

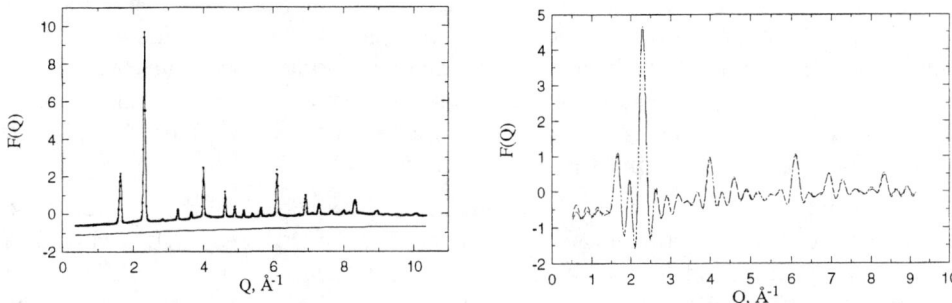

FIGURE 1. Left: experimentally measured $F(Q)$ for ND_4Cl [7]. Right: convoluted $F(Q)$ (solid curve) and RMC fit (dash).

An alternative was provided by the development of the MCGR method for crystalline systems. This is a modification of the MCGOFR method originally developed by Soper [8] and has been described in detail elsewhere [9]. Essentially it is a one dimensional equivalent of RMC where $g(r)$ is generated as a numerical function and its transform is fitted to the measured $F(Q)$. Since the function is only numerical it can be defined over as wide an r range as wished, thus avoiding truncation in the transform. Only the lower r part of the complete $g(r)$ is subsequently used for RMC modelling.

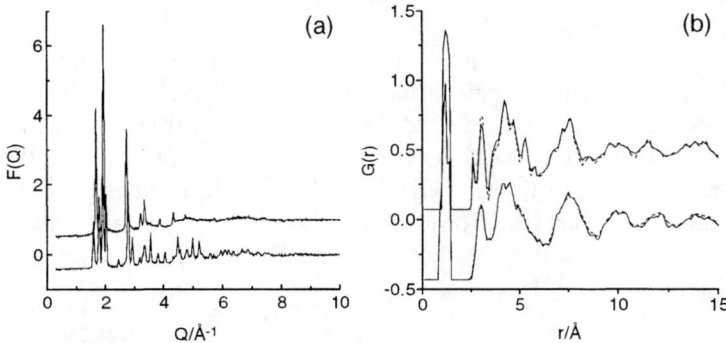

FIGURE 2. Left: $F(Q)$ for KCN at two temperatures (solid curves) and the MCGR fits (dash – the fits are so good that only the solid curves can be seen) [11]. Right: MCGR derived $g(r)$ (solid curve) and the RMC fits (dash).

The MCGR r range is effectively determined by the best resolution (narrowest peak) in the diffraction pattern. In order to make the transform sufficiently fast in practice it is necessary to pre-calculate all required values of $\sin Qr/Qr$ (eq. 6), which requires a lot of computer memory. For the data shown in figure 2, which are defined at approximately 10^3 Q points, it is appropriate to use an r range of about 200 Å, which with an r spacing of 0.05-0.1 Å (appropriate given the Q range) means 2-4×10^6 $Q \times r$ points. While in principle one could convolute the experimental resolution before the comparison of experimental and calculated structure factors, in practice this would mean an increase of an order of magnitude or more in the required r range and the calculation would then only be feasible on rather large computers. For high resolution pulsed source data which cover a wide Q range with different resolution functions for different angle detector banks, a real supercomputer would be needed. However a modification of the algorithm to significantly reduce the memory requirement is currently being tested [10].

This technique is not suitable when the sample has magnetic scattering. Because the magnetic moment is a vector quantity there is no direct equivalent to $g(r)$, which is a scalar quantity. Even small amounts of magnetic scattering can introduce significant distortions into the low r part of $g(r)$ and hence into the RMC model structure. In the end it is also preferable to make the RMC fit to a function which is as close as possible to the measured data in order to have the best control over the effect of errors. For this reason we have developed the RMCPOW method.

RMCPOW is a generalisation of the RMCX method originally developed for modelling single crystal diffuse scattering [12]. The principle can be described in the following way. Let us consider a simple crystal structure - for convenience we choose body centred cubic (bcc). The diffraction pattern for a perfect bcc lattice with one atom per primitive unit cell, at $T = 0$, is shown in figure 3. If T increases then the atoms start to vibrate about the perfect lattice sites. This leads to a reduction in the intensity of the Bragg peaks, which can be described in the first approximation in terms of the Debye-Waller factor (eq. 8). Where does the 'lost' Bragg peak intensity go? It ends up 'between' the Bragg peaks as the diffuse scattering. To illustrate this we have calculated the structure factor for a supercell consisting of 4×4×4 bcc unit cells. The calculation is made directly via eq. 5 and not via Fourier transform (eq. 6). If we only consider the 'average' unit cell within the supercell then we only obtain scattering at the same Q points as for the perfect bcc cell, with the intensities appropriately reduced because of the atomic msd. However if we consider all of the atomic positions in the supercell explicitly then we obtain scattering at all of the Bragg peaks of the supercell . Since the symmetry is only P1 the number of peaks is large. Because the atomic displacements are small relative to the original bcc lattice positions the strongest Bragg peaks are the subset which also belong to the bcc lattice.

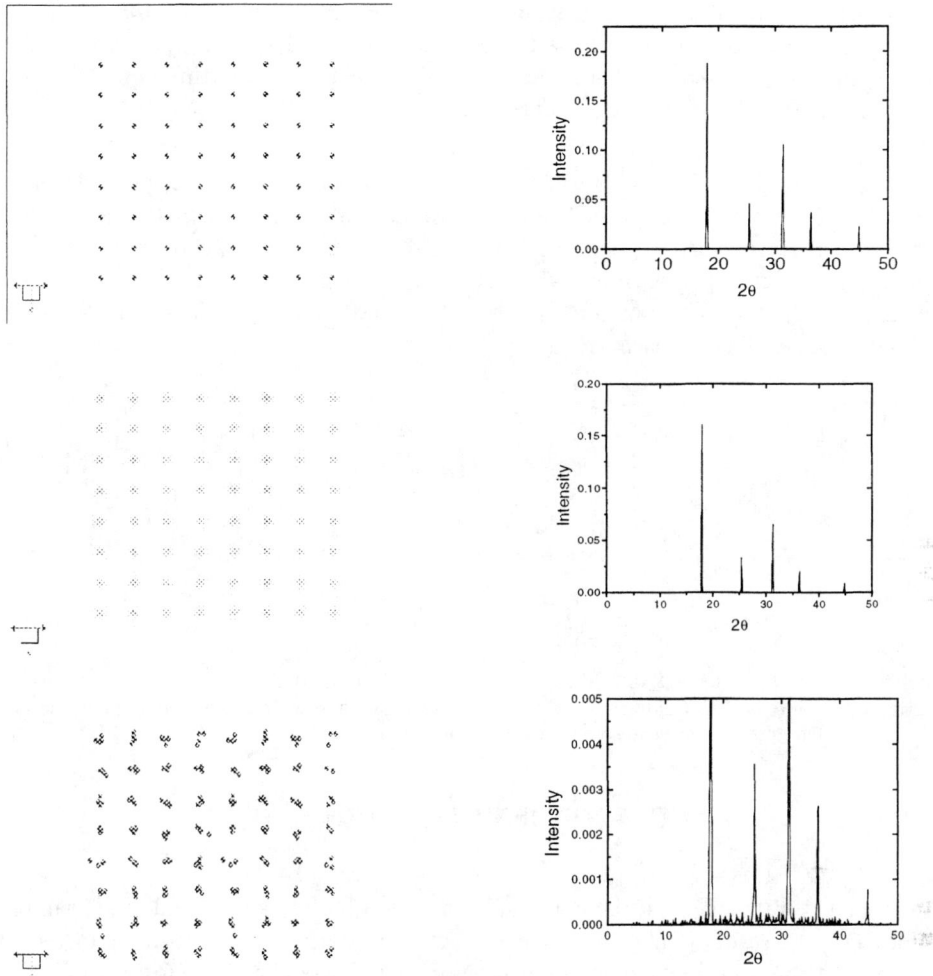

FIGURE 3. Model configurations of 4×4×4 bcc unit cells (left – shown in projection along a cube axis) and the corresponding structure factors (right). Top: perfect crystal, $F(Q) = S(Q)$. Centre: finite T crystal with time average atom positions and corresponding $S(Q)$. The atomic msd is represented by Gaussian blurring of the atom positions. Bottom: finite T crystal with instantaneous atom positions and corresponding $F(Q)$.

It is easy to see now that if we increase the size of the supercell the number of small Bragg peaks in between those of the original cell will increase until they become effectively a continuum - this is the diffuse scattering. If we can produce an RMC model (with periodic boundary conditions as normal) that is large enough then a direct calculation of the scattering due to this supercell give a reasonable approximation to both the Bragg and diffuse scattering. Because the pattern is calculated directly in Q space it can easily be convoluted with the experimental resolution function. If the sample is magnetic then a simultaneous model of the magnetic structure can be made, with the magnetic scattering being calculated according to similar principles (for details see [5,6]).

In practice for a reasonable size RMC model, say 4000 atoms, the supercell peaks are sufficiently numerous but not uniformly distributed in Q. This means that at low Q, where the peaks are sparse, it is necessary to smooth between them to obtain a good estimate of the diffuse scattering, while at high Q where they are numerous it is necessary to skip calculation of some of them to speed up the calculation. In figure 4 we show some results for $La_{0.8}Sr_{0.2}MnO_3$ [13].

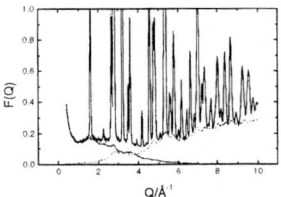 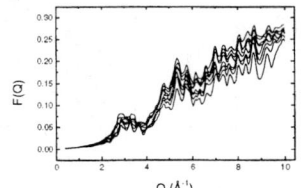 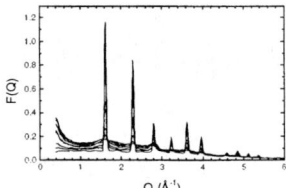

FIGURE 4. Left: $F(Q)$ for $La_{0.8}Sr_{0.2}MnO_3$ with RMCPOW fit and diffuse lattice and magnetic scattering contributions [13]. Centre: variation of lattice diffuse scattering contribution as a function of T. Right: variation of magnetic scattering contribution as a function of T.

FUTURE DEVELOPMENTS

In Rietveld refinement it is common to refine both instrumental and sample parameters, e.g. resolution, offset, scaling factor etc. In most cases it is faster in practice to do this with Rietveld refinement before RMC modelling, but for example the atomic displacements and the resolution parameters are correlated so parameter refinement should be included into RMC. It should be stressed at this point that RMCPOW is never going to compete with Rietveld refinement for the majority of crystal structure refinements – the computer time required is orders of magnitude larger.

Since RMCPOW is a generalisation of the RMC single crystal diffuse scattering code, RMCX, it is relatively straightforward to modify RMCPOW to allow simultaneous modelling of powder and single crystal data. It might be asked why powder data are used when single crystal data are available? In the case of diffuse scattering, single crystal data are usually only measured over a restricted number of planes (possibly only one) and at a restricted number of points, due to practical time limitations. It is also much more difficult to absolutely normalise single crystal data and to measure Bragg scattering under precisely the same conditions as diffuse scattering. While the single crystal data therefore provides three dimensional information it is relatively incomplete, so the powder data provide a valuable (and possibly necessary) constraint to the modelling. In common with the RMCA code for liquids and amorphous materials, it is also possible to extend RMCPOW to modelling of X-ray and EXAFS data, and to include chemical constraints.

ACKNOWLEDGEMENTS

This work has been supported by the Swedish Natural Sciences Research Council (NFR).

REFERENCES

1. Keen, D.A., Hayes, W., McGreevy, R.L., and Clausen, K.N., Phil. Mag. Lett. **61,** 349 (1990)
2. McGreevy, R.L., and Pusztai, L., Mol. Simulation. **1** 359 (1988)
3. Keen, D.A., and McGreevy, R.L., Nature **344** 423 (1990)
4. Nield, V.M., Keen, D.A., Hayes, W., and McGreevy, R.L., J. Phys. Cond. Matter **4** 6703 (1992)
5. Mellergård, A., and McGreevy, R.L., Acta Cryst. A in press (1998)
6. Mellergård, A., and McGreevy, R.L., J. Phys. Cond. Matter **10** 9401 (1998)
7. Belushkin, A.V., Kozlenko, D.P., McGreevy, R.L., Savenko, B.N., and Zetterström, P., Physica B submitted (1998)
8. Soper, A.K., in *Neutron Scattering Data Analysis*, ed. Johnson, M.W., IOP Conference Series **81** 71 (1986)
9. McGreevy, R.L., and Pusztai, L.,. Physica B **234-236** 357 (1997)
10. Zetterström , P., private communication (1998)
11. Karlsson, L., and McGreevy, R.L., Physica B **234-236** 100 (1997)
12. Nield, V.M., Keen, D.A., and McGreevy, R.L., Acta Cryst. A **51** 763 (1995)
13. Mellergård, A., McGreevy, R.L., and Eriksson, S.-G., private communication (1998)

Modelling the Thermal Expansion of Zeolites

J.D. Gale

Department of Chemistry.
Imperial College of Science, Technology and Medicine, South Kensington, SW7 2AY, U.K.

Abstract. Free energy minimisation, using analytical derivatives, has been applied to three different cubic zeolite structures including faujasite which experimentally has been demonstrated to show uniform negative thermal expansion over a wide temperature range. Both total and cell-only free energy minimisation have been examined for three different interatomic potential sets. For all models the former method is found to breakdown close to room temperature and therefore the cell-only approach is the method of preference.

INTRODUCTION

Zeolites are a family of aluminosilicate minerals that have found many applications as a consequence of their microporous structures where the channels are typically of dimensions comparable to small molecules. In addition, the presence of both Bronsted and Lewis acid sites leads to shape selective catalytic activity. However, less use has been made of the mechanical properties of such materials. Since the observation was made that zeolites can contract under heating they have attracted interest as possible components of zero thermal expansion ceramics. In order to be useful for applications of this kind it is ideal for a material to demonstrate uniform contraction along all crystallographic directions, rather than just one [1]. Recently it has been showed that the faujasite structure possesses this attribute over a wide range of temperature [2].

The aim of this study is to consider the application of theoretical methods to examine whether it is possible to reproduce the negative thermal expansion of the mineral faujasite and beyond this to study the temperature dependent structures of other cubic zeolite structures that may be possible candidates.

To date the majority of simulations of ionic materials that have utilised interatomic potentials have involved athermal minimisation of the lattice energy to find the local energy minimum. However, the temperature dependance of structure and properties can be introduced via one of two complementary approaches. In the high temperature regime, particular when solids are close to undergoing a continuous phase transition between two structures or ultimately melting, the method of preference is molecular dynamics as this is able to routinely handle strong anharmonicity [3]. However, at low temperatures molecular dynamics is invalid as it neglects the quantisation of the vibration levels leading to population of states which should not be accessible at the designated thermal energy. Strictly speaking, this approach is only appropriate once the Debye temperature has been reached which for many ceramics can be a long way above ambient conditions [4].

In the low temperature regime the technique of preference is to use lattice dynamics to evaluate the free energy through the use of statistical mechanics. The difficulty here is that the dependence of the free energy on the phonon density of states and in turn on the structure is in general complex. Only recently has it become possible to analytically determine the free energy gradients needed in order to find the minimum free energy structure at a given temperature without approximating the phonon density of states in some way. Earlier studies by Parker and co-workers [5] performed free energy minimisation of unit cells using numerically determined derivatives. However, using this approach it was impractical to optimise internal degrees of freedom with respect to the free energy as well as the cell parameters. Hence this gave rise to the form of the quasiharmonic approximation referred to as the Zero Static Internal Strain Approximation (ZSISA), in the notation of Taylor et al [6], as opposed to full free energy minimisation. With the advent of analytical derivatives, both algorithms can now be contrasted.

Here we present results on the application of this new approach for analytical free energy minimisation to a number of cubic zeolites to determine their thermal expansion properties based on a number of standard interatomic potential models for silicates. Also the results of both full and cell-only free energy minimisation will be compared for these materials.

METHODS

The underlying theory for analytical free energy minimisation has been recently developed by Kantorovich and applied to alkali halide crystals [7], [8]. Subsequently the method has been refined by Taylor et al [6] who have discussed many of the details of its implemenation. Hence only a brief summary of the salient features will be given here. More details concerning the particular implementation used here can be found elsewhere [9].

The Helmholtz free energy for an ordered system can be written as the sum of the static internal energy, U_{static}, that would be calculated in a conventional energy minimisation, the vibrational energy, U_{vib}, and the term arising from the vibrational entropy, S_{vib} :

$$A = U_{static} + U_{vib} - T.S_{vib} \qquad (1)$$

For convenience we can express the sum of the vibrational energy and entropy term together, due to the cancellation of a common term, as:

$$U_{vib} - T.S_{vib} = \sum_k \sum_m \left\{ 1/2 h\bar{\omega}_m(k) + k.T.\ln\left[1 - \exp\left(-\frac{h\omega_m(k)}{k.T}\right)\right]\right\} \quad (2)$$

Here the sum over K points is used to approximate the integral over the Brillouin zone of the phonon density of states. The vibrational frequencies at each K point are given by the square root of the eigenvalues of the dynamical matrix, which in turn are related to the phased second derivative matrix multiplied by vectors containing inverse the square root of the atomic masses:

$$\omega^2(k) = e^{-1}(k).D(k).e(k) \quad (3)$$

$$D_{\alpha\beta}^{ij}(k) = \frac{1}{(m_i.m_j)^{1/2}} \sum_l \left(\frac{\partial^2 U_{static}}{\partial\alpha\partial\beta}\right).\exp\left(\imath k(r_{jl} - r_{i0})\right) \quad (4)$$

The derivatives of the free energy with respect to structural parameters can be related to the derivatives of the eigenvalues or frequencies squared. Hence the main difficulty lies in calculating the derivatives of the eigenvalues. Through the application of perturbation theory, these derivatives can be related to derivatives of the elements of the dynamical matrix projected onto the eigenvectors of each phonon mode:

$$\left(\frac{\partial\omega^2}{\partial\epsilon}\right) = e_m(k).\left(\frac{\partial D(k)}{\partial\epsilon}\right).e_m(k) \quad (5)$$

The first derivatives of the dynamical matrix elements are just the third derivatives with respect to either three Cartesian coordinates, for internal degrees of freedom, or two Cartesian coordinates and the external strain in the case of the unit cell derivatives. Both must also be multiplied by the appropriate phase factor for the point in the Brillouin zone.

Having obtained the first derivatives of the free energy we need to be able to efficiently optimise the geometry of the system with respect to this quantity. Here we follow the approach of Taylor et al [6] which uses a Newton-Raphson method based on an approximate hessian matrix which is calculated from the static second derivatives only. This matrix is updated subsequently using the BFGS scheme which will, in principle, tend to correct for the missing vibration contribution over sufficient cycles. In practice, the static hessian is already a good approximation and so only a few a cycles of minimisation are required when starting from the statically optimised structure.

The above scheme generates both internal and external derivatives with respect to the free energy. However, for comparison we would also like to be able to perform calculations within the zero static internal stress approximation (ZSISA), in the notation of Taylor et al, as used previously in the numerical formulation. In this case the internal variables must be minimised with respect to the internal energy while only the strain variables are minimised with respect to the free energy. To achieve this we must first neglect the thermal contribution to the internal forces. However, there will also be a correction term arising for the strain derivatives associated with the fact that the internal energy must remain at its minimum point as the cell is strained [10]. This can be shown to be:

$$\left(\frac{dA}{d\epsilon}\right)_{ZSISA} = \frac{\partial A}{\partial \epsilon} - \frac{\partial^2 A}{\partial \epsilon \partial \alpha} \cdot \left(\frac{\partial^2 A}{\partial \alpha \partial \beta}\right)^{-1} \cdot \frac{\partial A}{\partial \beta} \qquad (6)$$

As we wish to avoid calculating the second derivatives with respect to the free energy due to the complexity and computational cost we can approximate the two second derivative matrices by their static only components. Because one matrix is multiplied by the inverse of the other there will be a significant cancellation of errors and this turns out to be a good approximation in practice.

All calculations in this study have been performed using the program GULP [11] which is currently able to calculate analytical free energy derivatives of 2-, 3- and 4-body forces. The symmetry features of the program are used to restrict geometry optimisations to the asymmetric unit variables, although at this stage no attempt is made to use symmetry to accelerate the calculation of the free energy derivatives. Previous results suggest that quite a substantial benefit should be gained from implementing this in future.

In the case of both faujasite and zeolite-A the size of the unit cell is large enough such that a single K point at 0.25 / 0.25 / 0.25 is sufficient to sample reciprocal space. However, the smaller unit cell of sodalite requires a grid of K points chosen according to the Monkhorst-Pack scheme and symmetrised according to the Patterson space group [12]. Here a mesh size of 6 x 6 x 6 was utilised so as to achieve the same degree of convergence as for the other two structures.

In order to perform the free energy minimisations we must have an underlying model for the energy of the material. Here we work primarily within the ionic model and assume that the dominant term is the Coulomb interaction between ions which is supplemented at short range by a Buckingham potential. Furthermore, we can include ionic polarisation via the shell model [13] which consists of a core coupled to a mass-less shell by a harmonic spring constant. There are already exist several different parameter sets for modelling silica polymorphs within this framework. Amongst the most widely used are the formal charge shell model potentials of Sanders et al [14], often used in the slightly modified form given by Jackson and Catlow (JC) [15], and the partial charge rigid ion model of van Beest et al (VB) [16]. In addition, we recently refitted the Sanders model to allow for the effect of the thermal expansion and the zero point energy term (FED) [9]. Also the three-body angle bending term was replaced in the new model by a Urey-Bradley strut potential as this is more computationally efficient with no apparent loss of accuracy for quartz. The parameters for all three potential sets can be found in the original references and all have been tested to examine their relative performance for thermal expansion predictions in this work.

TABLE 1. Calculated bulk properties of ZDDAY according to the three different potential models used

Property	FED	JC	VB
$a_{athermal}$ (Å)	24.3269	24.2262	24.7811
a_{300K} (Å)	24.3065	24.2201	24.7793
Bulk Modulus (GPa)	79.7	62.5	67.60
ϵ^0	2.70	2.65	1.40
ϵ^{inf}	1.45	1.50	1.00
ω_{max} (cm^{-1})	1156	1083	1238

RESULTS

The ideal choice of negative thermal expansion material, as previously mentioned, is one that contracts uniformly. This criterion is known to be satisfied by the zeolite faujasite. In essence, the structure consists of sodalite cages connected by double six-ring units to give an open three-dimensional channel structure with a limiting aperture consisting of a twelve-ring. The commonly used forms of this material are known as zeolite-X and zeolite-Y, depending on the aluminium content of the framework. However, it is also possible to prepare an almost pure silica form which is referred to as ZDDAY [17]. It is this latter form that we will be considering here to avoid the complications of the influence of the aluminium distribution and counter ions on the thermal properties. For comparison with ZDDAY we also consider here two other cubic zeolites which have a strong structural relationship to this material, sodalite and zeolite-A (again both as pure silica polymorphs). As for faujasite, both structures are constructed from sodalite cages which are connected via double four-ring units for zeolite-A and directly fused together for sodalite. This triad of related structures should make for an interesting comparison of the influence of the connectivity of the sodalite cages on the thermal expansion.

As mentioned previously, three different interatomic potentials have been tested here in the case of the faujasite structure. The optimised unit cell parameters given by each model are listed in Table 1 for both room temperature (300K) and as obtained from an athermal calculation. The Jackson-Catlow variant of the Sanders et al potential model gives the best agreement with the experimental value of 24.2576 Å [17] for the unique cell parameter at 298K, closely followed by the free-energy derived (FED) model of Gale. This indicates the limitations of transferability of the potentials from quartz where the latter model is the more accurate. The rigid ion model of Van Beest et al is by far the worst in this case, giving an error in the cell parameter of over half an Angstrom. Unfortunately there are no experimental measurements of the bulk mechanical or electrical properties of purely siliceous faujasite, to the best of our knowledge, and hence we can only assume that the agreement would be similar to that found for α-quartz.

The calculated thermal expansion coefficients for ZDDAY are given in Table 2 for all three potential models using the ZSISA approximation. Values are quoted based

TABLE 2. Thermal expansion coefficients for purely siliceous faujasite as calculated by three different potential models

Temperature Range (K)	FED	JC	VB
	($\times 10^{-6} K^{-1}$)		
10-300	-4.42	-4.03	-5.24
10-500	-4.93	-4.25	-6.11
500-800	-5.66	-4.07	-9.52

upon three temperature ranges, (1) 10K - 300K, (2) 10K - 500K and (3) 500K - 800K. Additionally, in the case of the free energy consistent potentials the thermal expansion coefficient up to 250 K has been calculated to be -7.25 x $10^{-6} K^{-1}$ based upon full minimisation with respect to the free energy of all degrees of freedom.

As was discovered previously for quartz [9], full free energy minimisation is found to break down just below room temperature - a finding that is independent of the interatomic potential model used. What occurs is that imaginary modes begin to appear which suggest a phase transition to a lower symmetry. However, this is an artefact of the calculation as it occurs for all framework silicates at about the same temperature. The reason for this can be understood as follows. The free energy is dominated at low temperatures by the low frequency modes and the way to minimise the free energy is therefore by forming soft modes. In the ZSISA approach this cannot happen because the internal degrees of freedom responsible for the creation of soft modes are only optimised with respect to internal energy. However, in total free energy minimisation the atomic positions are directly coupled to this quantity. The fundamental problem lies in the fact that the phonons are being calculated as the second derivatives of the internal energy while the atoms are at a minimium with respect to the free energy. This inconsistency rapidly makes anharmonicity very important with increasing temperature. Hence this issue must be addressed to make total free energy minimisation useful for structures with relatively facile modes of distortion available to them.

Interestingly the thermal expansion coefficients calculated for purely siliceous faujasite are more accurate according to the ZSISA approach than with full free energy minimisation. Hence the breakdown of the later is not a hinderance to free energy minimisation being a useful predictive tool. Both of the shell model potential show reasonable quantitative agreement with the experimentally determined thermal expansion coefficient of -4.2 $x10^{-6} K^{-1}$ [2] in the low temperature regime, while the rigid ion potential overestimates the degree of contraction. The calculated value for the thermal expansion coefficient based on the Jackson-Catlow potential is in good agreement with the earlier determined value of Couves et al [18] using numerical free energy minimisation.

All potential models show a tail off in negative thermal expansion close to absolute zero. However, the behaviour at high temperatures shows qualitative differences. The Jackson-Catlow potentials show the correct behaviour in that the thermal expansion is close to linear over a wide temperature range and becomes slightly less negative towards the upper end. Both of the other models show increased negative thermal expansion at high temperatures, which is not experimentally the case. In the case of the free energy derived potential this is probably due to the incorrect coupling of the Si-O bond potential and the O-Si-O angle bending potential, due to the use of a strut potential, for large distortions.

Now we turn to consider the application of free energy minimisation to two other zeolites which exhibit cubic structures, zeolite-A and sodalite. Unlike faujasite, these two materials have yet to be prepared in a purely siliceous form to the best of our knowledge. However, it is still interesting to examine whether all members of this closely related family of frameworks containing the sodalite cage all demonstrate negative thermal expansion. For this part of the study all calculations are performed using the ZSISA approach only.

Extension of the calculations to these two further polymorphs in fact turns out to be problematical. Even when neglecting the internal free energy derivative contributions, it is found that imaginary modes are readily induced during minimisation for both structures. The point at which the structures become unstable is now found to be much more sensitive to the potential model used, though not in a consistent fashion. For instance, the FED model gives a valid minimisation of the zeolite-A structure up to room temperature while the JC model fails close to zero, while for sodalite almost the reverse situation occurs. Consequently results will be presented for zeolite-A based upon the FED potentials, while for sodalite the JC parameters will be used.

The breakdown of the free energy minimisation of the sodalite structure occurs due to the presence of four imaginary modes, of which the first three are triply degenerate. Lowering of the symmetry to remove these modes suggests that, according to this potential model, there would be a cubic to rhombohedral transition for purely siliceous sodalite at relatively low temperatures. In reality, even if such a distortion were to be favourable the thermal motions would probably lead to a cubic average structure and thus no change would be observed. However, the key feature that emerges is that the thermal expansion for sodalite is positive and therefore is not so interesting in the context of this work.

The value of the cubic cell parameter of zeolite-A versus temperature, calculated using the free energy derived potentials, is shown in Figure 1. As in the case of faujasite the thermal expansion is negative and isotropic. However, the magnitude appears to be even larger with an average coefficient of $-10.7 \times 10^{-6} K^{-1}$ over the temperature range plotted - approximately twice that of faujasite. Unfortunately beyond 200 K, with all potential models, the structure develops imaginary modes, again indicating a possible dynamical rhombohedral distortion.

The reason why faujasite and zeolite-A show negative thermal expansion while sodalite doesn't can be readily understood along the lines proposed from Rigid Unit Mode theory [19]. According to this theory the tetrahedra can be viewed as essentially rigid with the critical low frequency vibrations involving the cooperative rotations of tetrahedra. In some structures there are such modes which lead to

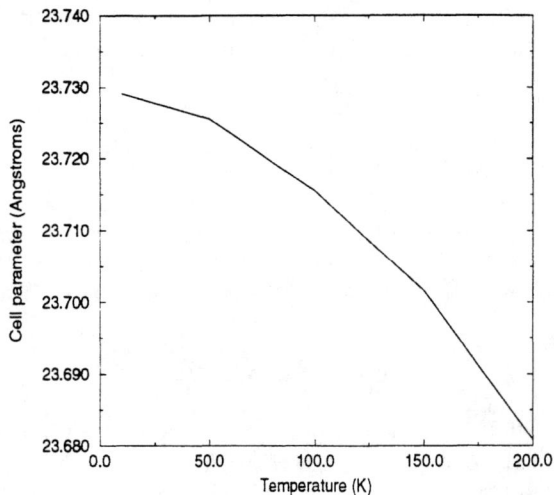

FIGURE 1. Cell parameter as a function of temperature for siliceous zeolite-A.

overall contraction of the unit cell when the units rotate. Similarly for the triad of structures considered here we can consider the sodalite cage as being a quasi-rigid unit and therefore it is the relative rotations of these units which lead to negative thermal expansion for faujasite and zeolite-A. However, in the case of sodalite the cages are directly fused via four-rings which removes the ability to rotate significantly. Hence, the positive thermal expansion of the bond lengths dominates. A challenge that remains for the future is to be able to explain the behaviour of a number of zeolites that show positive thermal expansion at low temperatures which turns to negative thermal expansion at higher temperatures [20].

CONCLUSIONS

Free energy minimisation within the quasiharmonic approximation is found to give good quantitative reproduction of the experimentally observed negative thermal expansion of purely siliceous faujasite, provided only the cell parameters are minimised with respect to the free energy and a shell model potential including three-body interactions is used. Complete minimisation of all structural parameters with respect to the free energy is found to fail close to room temperature regardless of the potential model used.

Purely siliceous zeolite-A is predicted to show even greater thermal expansion than the faujasitic equivalent and may be worth exploring experimentally in future. In contrast, the direct coupling of the sodalite cages in sodalite prevents the contraction mechanism from operating, leading to conventional positive thermal expansion.

ACKNOWLEDGEMENTS

I would to thank Lev Kantorovich, Neil Allan, Mark Taylor and co-workers for many useful discussions, as well as The Royal Society for the provision of a University Research Fellowship and EPSRC for computing facilities.

REFERENCES

1. Evans J.S.O., Mary T.A., Sleight A.W., Vogt T., *Science* **272**, 90 (1996).
2. Attfield M.P., Sleight A.W., *Chem. Commun.* 602 (1998).
3. Allen M.P., Tildesley, D.J., *Computer Simulation of Liquids*, Oxford:Clarendon (1987).
4. Dove M.T., *Introduction to Lattice Dynamics*, Cambridge University Press (1993).
5. Parker S.C., Price G.D., *Adv. Sol. State Chem.* **1**, 295 (1989).
6. Taylor M.B., Barrera G.D., Allan N.L., Barron T.H.K., *Phys. Rev. B* **56**, 14380 (1997).
7. Kantorovich L.N., *Phys. Rev. B* **51**, 3520 (1995).
8. Kantorovich L.N., *Phys. Rev. B* **51**, 3535 (1995).
9. Gale J.D., *J. Phys. Chem. B* **103**, 5423 (1998).
10. Allan N.L., Barron T.H.K., Bruno J.A.O., *J. Chem. Phys.* **105**, 8300 (1996).
11. Gale J.D., *JCS Faraday Trans.* **93**, 629 (1997).
12. Monkhorst H.J., Pack J.D., *Phys. Rev. B* **13**, 5188 (1976).
13. Dick B.G., Overhauser A.W. *Phys. Rev.* **112**, 90 (1958).
14. Sanders M.J., Leslie M., Catlow C.R.A., *JCS Chem. Commun.* 1271 (1984).
15. Jackson R.A., Catlow C.R.A., *Mol. Simul.* **1**, 207 (1988)
16. van Beest B.W.H., Kramer G.J., van Santen R.A., *Phys. Rev. Lett.* **64**, 1955 (1990).
17. Hriljac J.A., Eddy M.M., Cheetham A.K., Donohue J.A., Ray G.J., *J. Solid. St. Chem.* **106**, 66 (1993).
18. Couves J.W., Jones R.H., Parker S.C., Tschaufeser P., Catlow C.R.A., *J. Phys. Cond. Matter* **5**, L329 (1993).
19. Hammonds K.D., Heine V., Dove M.T., *Phase Trans. B* **61**, 155 (1997).
20. Park S.H., Grosse Kunstleve R.-W., Graetsch H., Gies H., *Studies in Surf. Sci. and Catal.* **105**, 1989 (1997).

Multiple Scattering Effects in Deep Inelastic Neutron Scattering experiments

J.Dawidowski, J.J. Blostein and J.R. Granada

Comisión Nacional de Energía Atómica and CONICET, Centro Atómico Bariloche, (8400) San Carlos de Bariloche, Río Negro, Argentina

Abstract. We present a correction procedure to account for multiple scattering effects in Deep Inelastic Neutron Scattering experiments in molecular systems, based on a Monte Carlo simulation in which the energy-transfers are described with a synthetic model. Experimental results in polyethylene samples of different thicknesses performed at the Bariloche LINAC are presented along with numerical simulations that show a good agreement between experiment and calculations. It is shown that multiple scattering effects may cause an important non-symmetric effect on the observed peaks.

INTRODUCTION

The correction of multiple scattering effects in neutron scattering experiments has recently been the focus of new studies (1,2), in which it was emphasized that specific numeric algorithms for each experimental situation are needed, provided that a detailed description of the experimental setup must be included in the calculations. In many cases, experimental data obtained in a given experiment, is not enough information to undertake a complete simulational correction procedure, either because the experiment is deviced to explore some integral magnitude or (even in the most detailed kind of experiments), because data out of the limits of the explored dynamic range are needed. It is also worth noticing, that in every case, a knowledge of the total cross section as a function of the neutron energy is needed to calculate the mean-free-path at each step, a magnitude which is commonly overlooked in the present day neutron activities.

Multiple scattering, attenuation and empty-can effects are magnitudes that usually will significantly affect a neutron scattering experiment, even if it is performed in a high intensity neutron source, where the use of small samples can be afforded. This can be so if some low-signal portion of the sample spectrum is to be explored.

In this work we analyze the Multiple Scattering effects on Deep Inelastic Neutron Scattering (DINS) experiments, a technique especially suited for probing the momentum distributions of atoms (3). We present a Monte Carlo code specialized in this kind of experiments, which uses a Synthetic Model (4) for the case of molecular systems, and a gas model with an effective temperature (5) for solids. The results are compared with experimental data on simple systems of different sizes to vary the amount of multiple scattering and attenuation.

EXPERIMENTAL SETUP

The experiments were performed at the Bariloche electron LINAC (Argentina). Neutrons were obtained through collision of 25 MeV electrons with a lead target, at a rate of 100 pulses per second. A 4-cm thick polyethylene moderator was employed, and a cadmium sheet was inserted in the neutron beam to select the epithermal part of the spectrum. The sample was placed at 490 cm from the moderator, and neutrons were detected at an angle of 55° by six ^{3}He detectors (10 atm. pressure, 1 in. diameter and 6 in. active length) placed in a circular-corona geometry at 29 cm from the sample. An Indium foil 0.05 mm thick was used as resonant filter, and ´foil in´ and ´foil out´ runs were performed every 30 minutes and subtracted after monitor normalization.

MULTIPLE SCATTERING CALCULATIONS

The basic description of the Monte Carlo algorithm is based on Sears´ formalism (6) and was presented elsewhere (2), while the numerical method is similar to that proposed by Copley (7) many years ago, so only a short description will be given here. Neutron histories are followed individually, and the flight path at each step is determined by the distribution defined by the mean free path of the materials (sample and container) that the neutron has to traverse at the current energy, which means that a detailed knowledge of the total cross section is needed. The weight is diminished by the fraction of scattered and absorbed neutrons in the current path, and a history finishes when the weight drops under a predetermined value. At each step, the contribution to each time-of-flight channel of the fraction of neutrons absorbed by the resonant filter is computed, and new energies and flight directions are assigned randomly.

The assignment of energies and flight directions at each step is made employing the Synthetic Model (4) in the case of a molecular system, and a gas with an effective temperature (5) based on a Debye model for solids. The Synthetic Model describes the neutron-molecule interaction through a series of Einstein oscillators and the definition of effective masses, temperatures and vibrational factors which are functions of the neutron energy. Its expressions are analytic, which allows fast calculations in the Monte Carlo algorithm.

RESULTS

In Fig. 1 we show the results corresponding to a solid and a molecular system, i.e. graphite and heavy water. Graphite sample was 3.5 cm diameter and 1 cm thick, while heavy water was contained in an aluminum can 4 cm inner diameter and 1.5 cm thick, and 1 mm wall thickness. In the graphite sample an asymmetry in the lower time-of-flight part of the peak is apparent. This can be explained with the help of the Monte Carlo simulation, as a contribution of the multiple scattering component, corresponding to high-energy neutrons that after many collisions, reach at the energy of the resonant Indium filter (1.457 eV). In the case of heavy water we observe the deuterium and the oxygen peaks and again a multiple scattering component which causes a deformation towards low times of flight. The rest of the deformation in that direction corresponds to the contribution of the contents of 4% hydrogen of our sample.

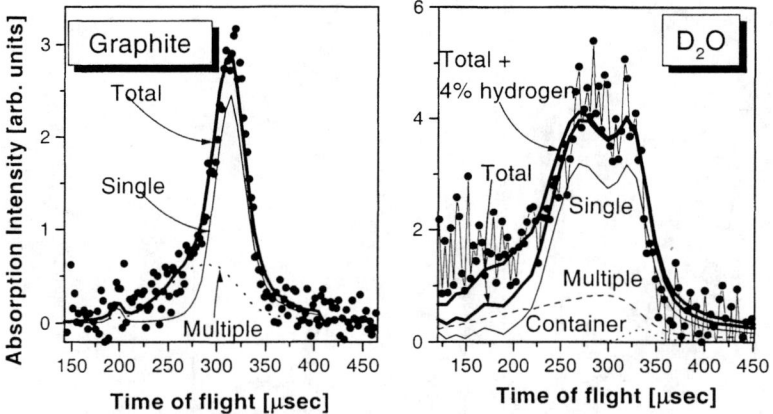

FIGURE 1. Experimental data obtained for graphite and heavy water (dots) compared with the Monte Carlo calculations, where the contribution of single and multiple scattering is shown. In the case of D_2O a 4% of contribution of hydrogen was added.

In Fig. 2 we show our results on samples of polyethylene of different thicknesses, performed to provide large amounts of multiple scattering to assess the calculation procedure presented here. Even for the thinnest sample (for which 72% of the neutrons are transmitted) the multiple scattering effects are apparent due to the lateral dimension of the sample (a coin shape 3-cm diameter).

CONCLUSIONS

We presented Monte Carlo procedure that describes satisfactorily multiple scattering effects in DINS experiments. These effects are shown to be important for the peak shapes, especially in their lower time-of-flight region. We analyzed experiments performed on a low-intensity source, where sample sizes cannot be reduced at will, in order to keep a good signal-to-noise ratio, however this corrections cannot be overlooked in experiments performed with small samples with high-intensity sources if low signal regions are to be analyzed. A complete description of the experimental setup had to be included, taking special care of the detector efficiency as a function of neutron energy.

The nucleus of this procedure was applied to different experimental situations with good success. However, a careful examination of each experimental situation is imperative for its application to new configurations.

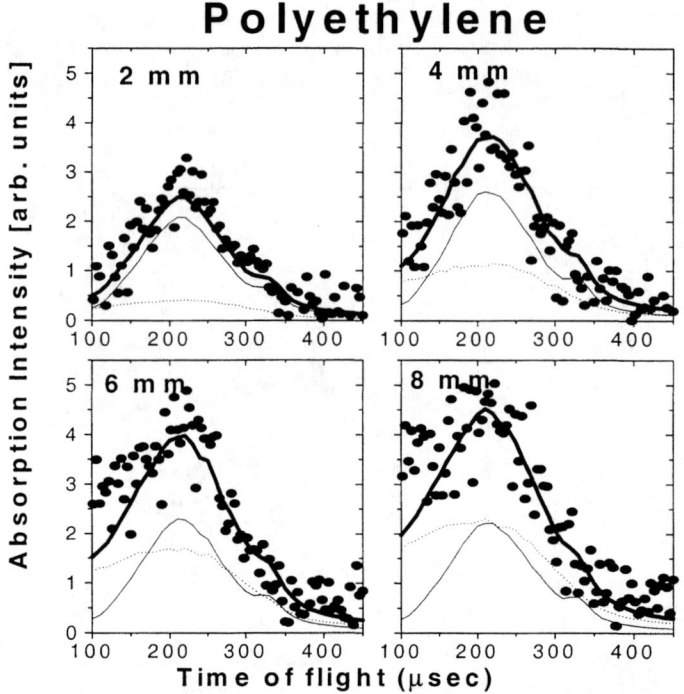

FIGURE 2. Experimental data on different polyethylene samples of different thicknesses. Dotted line indicates multiple scattering contribution, full thin line is single scattering, while the thick line is the total scattering.

ACKNOWLEDGEMENTS

The authors wish to acknowledge Fundación Antorchas (Argentina) for financial support of this work

REFERENCES

1. J. Dawidowski, G. J. Cuello and J. R. Granada, *Nucl. Instrum. Methods Phys. Res. B* **82**, 459 (1993).
2. J. Dawidowski, F.J. Bermejo and J.R. Granada, *Phys. Rev. B* **58**, 706 (1998).
3. J. Mayers, *Phys. Rev. B* **30**, 44 (1984).
4. J. R. Granada, *Phys. Rev. B* **31**, 4167 (1985).
5. J. R. Granada, *Z. Naturforsch.* **39a**, 1160 (1984).
6. V. F. Sears, *Adv. Phys.* **24**, 1 (1975).
7. J. R. D. Copley, *Comput. Phys. Commun.* **7**, 289 (1974).

A Monte Carlo Tool for Simulations of Neutron Scattering Instruments

L.L. Daemen, P.A. Seeger, T.G. Thelliez, and R.P. Hjelm

Manuel Lujan, Jr. Neutron Scattering Center
Los Alamos National Laboratory
Los Alamos, New Mexico 87545

Abstract. A brief description of the Neutron Instrument Simulation Package (NISP) is given, together with an example of its use to calculate the resolution function of an inelastic spectrometer.

WHY NISP?

The design of neutron scattering instruments and the exploration of novel techniques in neutron scattering are activities that have recently picked up pace. The trend is expected to continue with the recent endorsement by the U.S. Congress of the Spallation Neutron Source (SNS) project and with U.S. Department of Energy support for a number of pulsed or continuous sources enhancement projects at U.S. National Laboratories. Meanwhile, the European Spallation Source project is well underway, Germany is building a new research reactor in Munich, and the Paul Scherrer Institut source is now in production mode. Japan is also planning a 5 MW spallation source. Given the cost of these sources, carefully choosing and optimizing the performance of their associated suite of instruments is highly desirable. This task is less costly and more efficient than optimizing the neutron production target itself. At pulsed sources, however, the neutron pulse emitted by a moderator has a complex dependence on time, neutron energy, and direction of emission, all of which directly affect the performance of a given instrument. This greatly complicates the task of the instrument designer who now needs design tools beyond the traditional ray-tracing techniques of conventional optics. Even at continuous sources, there are many aspects of instrument design and optimization that cannot be addressed via ray-tracing techniques.

Analog Monte Carlo techniques provide a powerful complement to analytical design tools such as ray-tracing methods. Monte Carlo consists in sampling at random, but in a statistically representative manner, the phase space of the moderator to select the characteristics of a neutron. This neutron is then transported down the instrument axis and undergoes along the way all the interactions an actual neutron would feel in the real instrument until is either lost or detected. The interactions in such a history can be deterministic or statistical models of the relevant physical processes. The final answer is obtained by combining the results of many histories. While this approach has been used in the past to simulate isolated optical components or, less frequently, entire instruments, the associated codes were very much *ad hoc*. They lacked generality, took an excessive amount of time to program, and required advanced knowledge of Monte Carlo techniques.

The main purpose of our Neutron Instrument Simulation Package (NISP) was to overcome these shortcomings by providing the general user with little or no experience with Monte Carlo techniques a general, easy-to-use tool, simulation tool. This tool would allow the user to produce or modify a computer model of an instrument and simulate its performance on widely available computer platforms. While the original code was written by M. Johnson at the Rutherford-Appleton Laboratory in 1984 (1), the effort to make the code general started in 1994 at Los Alamos National Laboratory and produced a package that consists of three components (2):

MCLIB: a library of Fortran subroutines that deal with elementary tasks such a geometry representation, neutron transport, and contains all the models for optical elements. Associated with this library is a set of source files representing more than 25 different types of moderators, and a set of scattering kernels for sample simulation.

MC_RUN: a Monte Carlo engine that runs the Monte Carlo simulation itself and produces a series of output file with detector output and general information about the outcome of the simulation.

MC_Web: A WWW-based application that allows the user to set up the instrument geometry interactively without having to learn any of the data structures of MCLIB.

NISP has been used extensively during the past few years to design instrumentation for a number of LANSCE-based projects including the Long-Pulse Spallation Source (LPSS) project and the LANSCE enhancement project (currently underway). NISP has many uses beyond instrumentation design. It can be used to optimize optical components or entire instruments, analyze the performance of existing instruments, test novel ideas and concepts for components or instruments, verify design specifications, analyze experimental results, plan experiments on existing instruments, and teach neutron scattering techniques. NISP is freely available at http://bayberry.lanl.gov/lansce/Welcome.html.

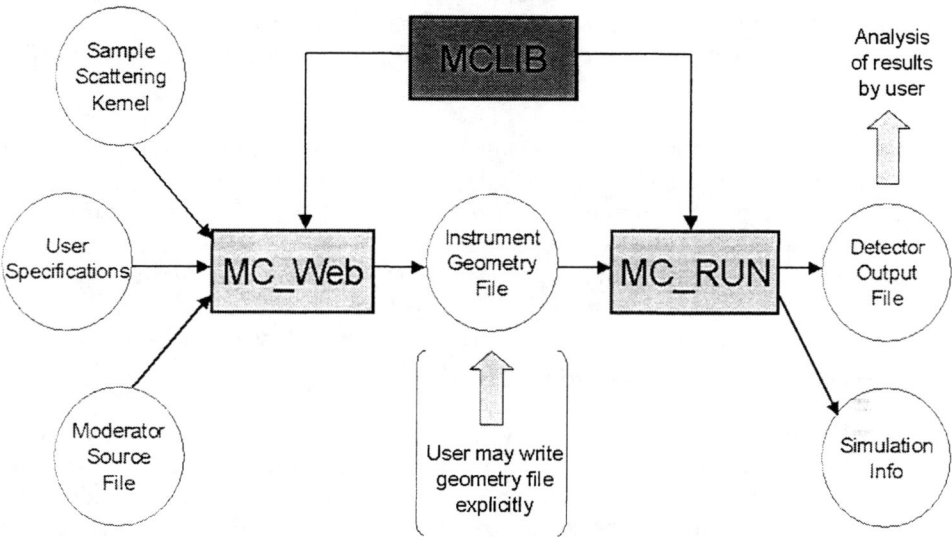

FIGURE 1. Simulation flow chart.

A typical simulation involves the generation of a computer model of the instrument or portion of instrument to be simulated. The user is assisted by MC_Web in this endeavor. The result is the so-called geometry file which contains all the information related to the instrument and necessary for the simulation. This file serves as an input file for the Monte Carlo engine, MC_RUN. The output produced by MC_RUN is essentially an ASCII file with 1D, 2D, or 3D histograms for each detector and a file with information about the course followed by the simulation, error messages, various summaries, etc... Both MC_Web and MC_RUN make use of the element definitions in the MCLIB library, Figure 1.

DATA ANALYSIS WITH NISP

An important step in neutron scattering data analysis consists in estimating the impact of instrument resolution on the data set. Instrumental resolution must be taken into account either by correcting the raw data set or is factored in at some stage of the analysis. Unfortunately, instrument resolution is not always easily determined. In the case of spectrometers for inelastic scattering, the elastic resolution function can be determined experimentally (usually with a strong incoherent scatterer such as vanadium). But at non zero energy transfer, and particularly at large energy transfers, it is difficult to measure a resolution function at a given momentum and energy transfer because few substances actually provide the experimentalist with sufficiently sharp energy levels to allow the determination of instrumental resolution.

In the absence of an efficient method for determining instrumental resolution experimentally, one may turn to analytical estimation or a Monte Carlo simulation. The former is possible only in the simplest cases and usually at the cost of drastic simplifications. The latter is ideal in that Monte Carlo can integrate over the large number of parameters defining the components of a neutron scattering instrument and take into account component imperfections that are not easily handled analytically. Furthermore, by using "ideal" samples with sharp energy levels, one can calculate a resolution function anywhere in (q,ω) space.

NISP is particularly well suited for this exercise. The user can either use an idealized sample to calculate one or more resolution functions, or he can use his sample model to generate a sample scattering kernel for the Monte Carlo simulation and simulate the actual experiment in an attempt to reproduce the experimental results, Figure 2.

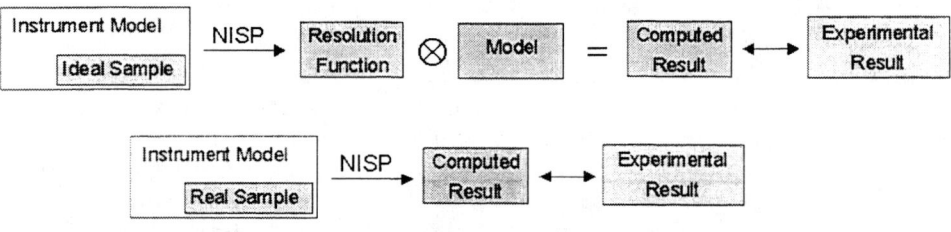

FIGURE 2. Two ways in which NISP can be used to assist in data analysis.

NISP is also useful to verify or validate experimental results, plan an experiment, and reconcile data sets obtained on two similar (but different) instruments.

EXAMPLE

One example will serve to illustrate the usefulness of NISP to calculate resolution functions. HERMES is a backscattering spectrometer for quasi-elastic and low-energy inelastic studies of materials. It is one of several new instruments to be built by the LANSCE enhancement project. As such, HERMES will have to share a partially coupled liquid hydrogen moderator with two other diffractometers. The partial coupling of the moderator results in an increase of the (time-integrated) intensity of the neutron pulse via the creation of a tail to the pulse that extents to long times. The effect of this tail on the quasi-elastic line is quite dramatic and is likely to complicate the data analysis, Figure 3. Clearly, assessing the effect of switching from a decoupled to a coupled moderator would have been a difficult task for analytical techniques.

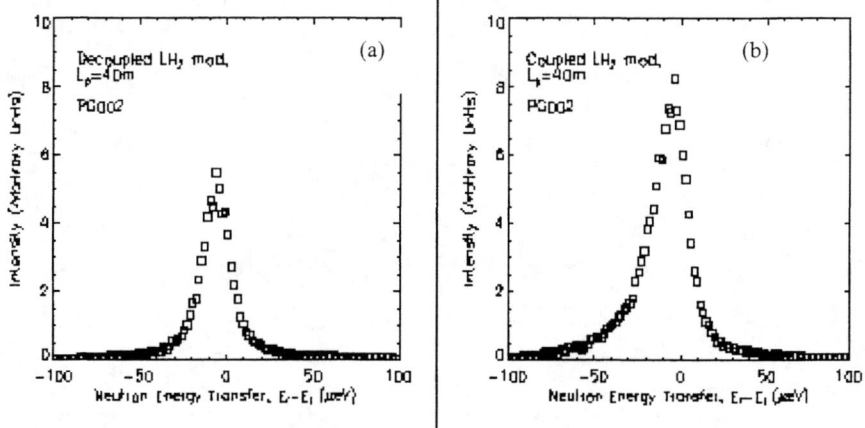

FIGURE 3. Elastic resolution function for HERMES (pyrolytic graphite 002 reflection) (a) on a decoupled liquid hydrogen moderator (decay constant of neutron pulse, $\tau=85$ µs), and (b) on a coupled liquid hydrogen moderator (decay constants of the neutron pulse, $\tau_1=160$ µs and $\tau_2=400$ µs).

NISP is designed to allow users to add new optical elements and new features to MCLIB, thus making NISP a highly flexible tool for the designer. The development of NISP parallels that followed extraordinarily successfully by the SHADOW code –the analog of NISP for x-ray instrumentation at synchrotron sources- now in use by thousands of users and the recognized authoritative tool in the field. This is our ultimate goal for NISP.

REFERENCES

(1) M.W. Johnson and C. Stephanou, "MCLIB: A Library of Monte Carlo Subroutines for Neutron Scattering Problems," RAL Report RL-78-090; M.W. Johnson, "MCGUIDE: A Thermal Neutron Guide Simulation Program," RAL Report RL-80-065.

(2) P.A. Seeger, "The MCLIB Library: Monte Carlo Simulation of Neutron Scattering Instruments,", Proceedings of the Thirteenth Meeting of the International Collaboration on Advanced Neutron Sources (ICANS XIII) held at the Paul Scherrer Institut October 11-14, 1995, G.S. Bauer and R. Bercher, Eds., PSI-Proceedings 95-01, p. 194. An updated version of this document , a tutorial, and additional documentation can be found at http://bayberry.lanl.gov/lansce/Welcome.html.

New Procedure For Multiple Scattering Correction

Margarita Russina and Ferenc Mezei

Hahn-Meitner-Institute Berlin
Glienicker Str.100
14109 Berlin
Germany

and

Los Alamos National Laboratory
MS H805

Los Alamos NM87545
U.S.A

We present a new approach to multiple scattering correction, based on a combination of taking experimental data at several incident neutron wavelengths and an iterative mumerical evaluation of multiple scattering contributions. It allowed us to model independently determine the low intensity part of dynamic structure factor in a typical glass $Ca_{0.4}K_{0.6}(NO_3)_{1.4}$. The new method makes the range of small momentum transfer in liquid and amorphous systems accessible for neutron inelastic scattering investigation and opens up new opportunities for the study of dynamic processes in these systems.

INTRODUCTION

The mystery of the glass transition is one of the hotly debated subjects in solid state physics since a long time. The similarity in the structure of glasses and liquids gives reason for the assumption that the transition between the glass and the liquid states is governed by dramatical changes in microscopic dynamics. The key to understanding of such processes lies in the intermediate range length scale (fig.1). This scale extends to several nearest neighbor atomic distance, thus collective and local atomic motions can be better distinguished here than at the scale of the short-range order.

A most powerful tool for studying the glass transition is neutron scattering. Contrary to other methods, neutron scattering provides simultaneous and direct information about dynamic processes on microscopic time and length scales. The reverse side of medal are the low incident flux of neutrons (in comparison to e.g. light scattering) and complex data treatment. One of a most nasty and difficult to handle problems of data correction procedures is the correction for multiple scattering. Multiple scattering can be made negligible by keeping the size of the scattering sample small. However, when dealin

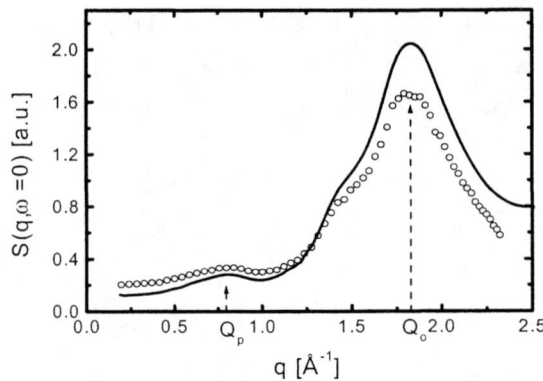

FIGURE 1. Ideal (solid line) and calculated "as measured" (circles) structure factor of CKN. The latter includes multiple scattering. Pointers indicate the intermediate and short range order, corresponding to Q_p and Q_0 respectively. The assumed sample was flat with thickness of 0.3 cm, the neutron wavelength 5 Å. Besides multiple scattering the calculation takes into account self screening and absorption.

with low intensities like a dynamical structure factor in glasses on the scale of intermediate range order (fig.1) any reduction of sample sizes leads to fatally small scattered intensities. Another common way, believed to be successful in the struggle against multiple scattering is the use of cadmium plates, installed horizontally in the sample. In order to check this method we calculated the inelastic scattered intensities for a typical glass $Ca_{0.4}K_{0.6}(NO_3)_{1.4}$ (CKN) with cadmium plates and without (fig.2). Our results show indeed a reduction of multiple scattering, but this contribution remains still much too high to allow us unambiguous interpretation of data.

After all, our conclusion is that there is no other choice but to live with multiple scattering and to consider it in the course of data evaluation. The effects of multiple scattering can not be removed from the spectra by subtraction or similar like the scattering on a sample cell. To estimate the contribution of multiple scattering correctly and reliably exact simulation of the scattering processes is needed and the quality of these simulations defines the quality of results. One general problem here is the knowledge of the scattering function $S(Q,\omega)$. The probability of neutrons to be scattered with certain energy and momentum transfer, the dynamical structure factor $S(Q,\omega)$, has to be known in the scattering angle range between 0 and 180 degrees. However, the whole angular range is not accessible experimentally. One has either to use some approximation and/or extrapolation or to use a theoretical dynamical structure factor, proposed by a model. This latter choice, however, leads to model dependent results. We developed a new model independent method of multiple scattering correction by using only experimentally determined $S(Q,\omega)$ [2],[3]. The new approach allows us to evaluate high precisely the scattering function, in particular

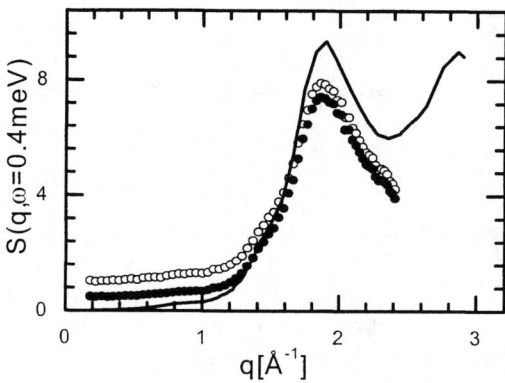

FIGURE 2. Theoretical dynamic structure factor $q^2S(q)$ of CKN (solid line) and calculated "as measured" spectra for the sample with (solid circles) and without Cd plates (open circles). The thickness of the sample is equal to 0.3 cm, distance between horizontal Cd plates is 0.3 cm.

its low momentum transfer part. The key of the method is to vary the multiple scattering contribution by making experiments at several incident neutron wavelengths and to combine these results with the help of simulation calculations which take into account multiple scattering and similar effects as exactly as possible. In the next chapter the new approach will be presented and demonstrated on the example of an amorphous system. Our sample is a CKN glass, which represents one of the most studied model systems in glass physics. The thickness of sample was equal to 0.3 cm. The experimental data were collected on the time-of-flight spectrometer NEAT in Berlin.

MULTIPLE SCATTERING CALCULATIONS

Similar to common methods [5],[6] our procedure to correct the multiple scattering (MS) consists of exact simulations of scattering processes and iterative comparison of the simulation results with experimental data until convergence is achieved. After every iteration step the model structure factor $S(Q,\omega)$ is modified and used as input for the next simulation. At the end the experimental spectra were corrected by the spectra difference between the simulated (multiple plus single scattering) and the model for $S(Q,\omega)$ used in the simulation. During the simulation procedure the history of every neutron is followed individually using Monte-Carlo techniques. The program starts with randomly setting up a neutron flight path $l_i=l_i(r_i,\Theta_i,\varphi_i)$, where r_i,Θ_i,φ_i are the polar coordinates. The probability for the neutron to be affected by the sample (absorbed or scattered) is characterized by a mean free path R_{tot}, experimentally determined by the measurement of the transmission coefficient of the neutrons with wavelength λ_i trough the sample.

$$p_i = \exp(l_i / R_{tot}) \quad (1)$$

The probability to be scattered is given by the mean scattering range R_{sc}:

$$p_i^{sc} = \exp(l_i / R_{sc}) \quad (2)$$

If the act of scattering takes place, the neutrons are scattered under a random angle Θ and energy change E and the probability of this process is given by the dynamical structure factor $S(Q,\omega)$. The next loop begins with setting up the new flight path of the scattered neutron. The neutrons, leaving the sample after one, two or three scattering events were tallied in predetermined scattering angle and energy channels Θ_d, E_d.

The novelty of our method is not contained in the simulation code but in the way of using this tool. The calculation procedure requires the exact knowledge of the mean free path R_{tot} and scattering range R_{sc}, which are defined by the macroscopic total cross section Σ and macroscopic scattering cross section Σ_{sc}, respectively. Following Sears [1], the double differential macroscopic cross section Σ_{sc} is proportional to an integral of the dynamical scattering factors over the range kinematically allowed in the

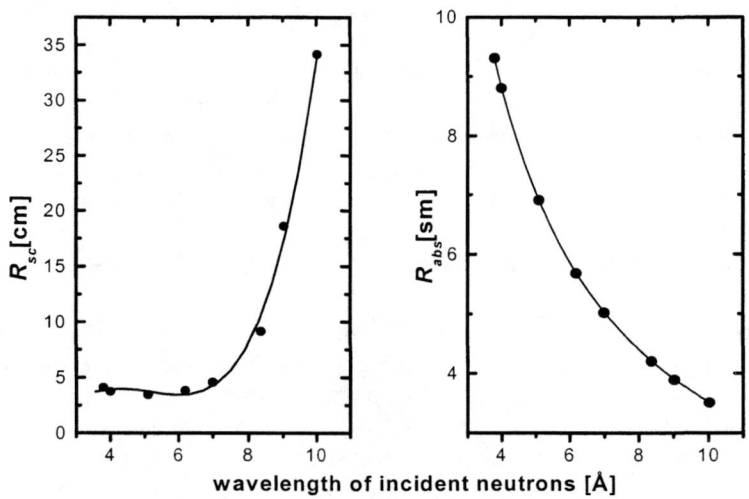

FIGURE 3. The neutron scattering range R_{sc} (a) and absorption range R_{abs} (b) of CKN. The absorption range was calculated by using the literature data. The values of the scattering range were then evaluated from the measured transmission coefficient at different wavelengths of incoming neutrons. The lines are guides to the eye.

experiment (for elastic scattering the experimentally covered kinematic range of momentum transfer Q is equal to $4\pi/\lambda_i$). Thus, by changing the wavelength of incident neutrons we change both the kinematic Q range and the macroscopic cross section Σ, and consequently, we are changing the contribution of multiple scattering to the spectra. We normalized the model scattering function used in the computation so that it corresponds to the experimentally determined R_{sc}. We measured the transmission of our sample as a function of the wavelength of the incident neutrons. Our results (fig.3a and 3b) show in fact the variation of neutron scattering range R_{sc} with λ_i. At long wave length R_{sc} is very large, because the experimentally accessible part of the dynamical structure factor has low values (Q<1.25 at λ_i=10 Å), i.e. low probability for a neutron to be scattered. The minimum of R_{sc} is at 6.2 Å, due to the main peak in the structure factor, which is reached at this wavelength.

At shorter wavelengths (λ< 6Å) the maximum in $S(Q,\omega)$ corresponds to a smaller range in scattering angle and the scattering probability becomes slightly smaller. An important implication of this observation is that by changing the wavelength of incident neutrons we change the probability of neutrons to be scattered, i.e. we can tune the multiple scattering contributions in the measured spectra.

The spectra of CKN-glass, obtained at three different wavelengths of incoming neutrons (fig.4.) demonstrate a clear difference at low Q's (i.e. the low scattering angles) caused by multiple scattering, which becomes most noticeable in this region. The spectra at 8.5 Å contain the smallest amount of multiple scattering in agreement with the analysis

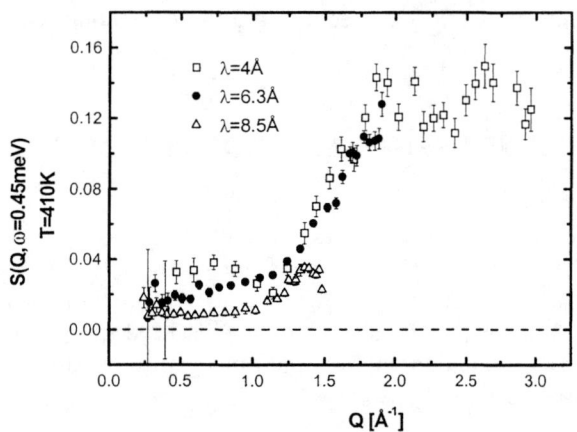

FIGURE 4. The spectra of CKN in supercooled liquid state, as measured at three different wavelengths of incoming neutrons. Differences between the spectra in the range of low Q's are due to the multiple scattering effect [2]. The points near Q=1.25Å^{-1} at λ_i=4Å and Q>1.3 at λ_i=8.5Å are strongly influenced by the absorption and self-screening in the direction of the flat sample plane.

above. The very high MS contribution in spectra at 4Å can be explained by additional multiple scattering processes from the sample environment. At this wavelength the Bragg scattering of Aluminum, contained in the cryofurnace and the sample cell, contributes very strongly to MS in the inelastic spectra of our sample. The MS processes from the spectrometer can be reduced experimentally by using an oscillating collimator around the cryofurnace, which was not available during our experiments. The elastic energy resolution was kept approximately constant over all wavelengths used, which allow us to enhance neutron intensities by using reduced chopper speeds at longer wavelengths. Our idea to determine the multiple scattering relies on a combination of data, taken at a number of incoming neutron wavelengths. Using the short wavelength we get information in extended range of Q and explore the region of small Q's by using the data obtained at long wavelength of incident neutrons. $S(Q,\omega)$ created this way is used as input for simulations of scattering at different λ_i, and good agreement between the experimental and simulated spectra is achieved already after the third iteration step at each wavelength (fig.5). Note, that the part of multiple scattering, coming from the cryostat at 4 Å, cannot be evaluated by a simulation, because of lack of information about the cross section and exact geometry of scattering processes in the cryofurnace. Therefore, we have used in this case the common procedure of identifying of the multiple scattering with $Q \to 0$ limit of the measured apparent $S(Q,\omega)$. This, however, does not impede the precision of our scheme: information from the 4 Å data is only needed for the high intensity, high Q part of $S(Q,\omega)$, where the MS contributions are relatively small. Most of the information on the small intensity, small Q regime comes from the 8.5 Å. The absorption and self-screening corrections, which cannot be exactly performed independently of MS correction, are done simultaneously as part of MS correction procedure.

DISCUSSION AND CONCLUSIONS

A new method for calculation of multiple scattering allows us to determine very precisely these detrimental contributions to measured spectra. The results of our simulations (fig.5) show good agreement with the measured data at all three wavelengths used. It can be clearly seen that the measured spectra at 6.3 Å wavelength of incoming neutrons contain about 70% multiple scattering at Q<1Å, while at 8.5 Å the MS part is only 20-30%.

By combining of data and doing the simulation for differentwavelengths we can reliably establish the ideal single scattering function in a wide range of Q, whereby information at small Q's primary comes from data taken at the long wavelength and little affected by MS, while the short incoming neutrons wavelength data delivery information at in the range of small Q's (fig 6.) The corrected spectra at different wavelength incident neutrons show agreement with each other (in contrast to the measured spectra in fig.4) and represent the true dynamical structure factor $S(Q,\omega)$.

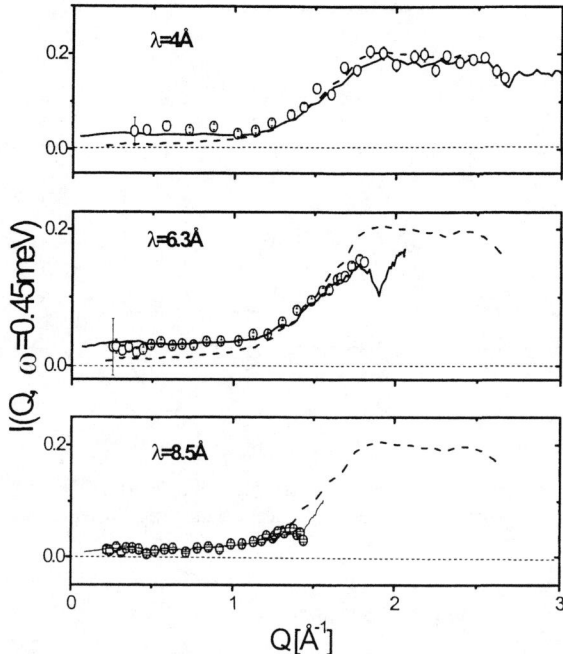

FIGURE 5. Simulated data (solid line) and measured spectra (open circles) at several wavelength incoming neutrons after third iteration. Dashed lines show the structure factor used as input. The dip is caused by enhanced self-screening in the direction of the plane of the flat sample.

The new approach, applied to amorphous systems, helps us to explain the microscopic processes, obtained near the glass transitions in a broad range of Q. The dynamical structure factor $S(Q,\omega)$ shows well pronounced deviations on the scale of intermediate range order from the $Q^2 S(Q,\omega)$ approximation, expected for sound like excitations [4] and reveals the first clear evidence for the existence of structural relaxation in fast dynamics regime near the glass transitions. In general, our method opens up a whole field of new opportunities of exploring processes in amorphous systems and liquids by making the range of low momentum transfer accessible to inelastic neutrons scattering study.

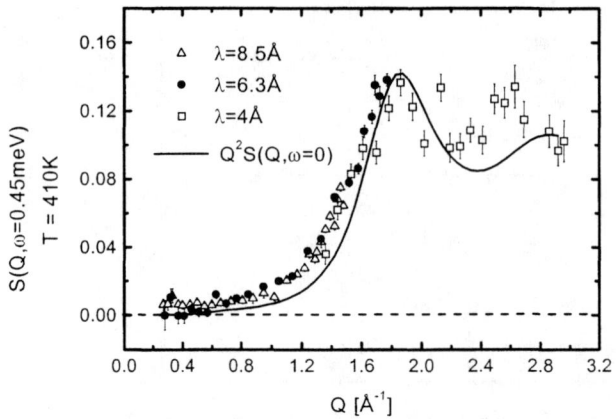

FIGURE 6. Multiple scattering corrected spectra at several incident neutrons wavelength. Clear deviations from $Q^2S(Q,\omega=0)$ low are seen in the region of intermediate range order.

REFERENCES

1. V.F. Sears, Adv.Phys. 24,1(1975)
2. M.Russina, F.Mezei, R.Lechner to be published.
3. F.Mezei, M.Russina, Workshop on "Non-Equilibrium Phenomena in Supercooled Fluids, Glasses and Amorphous Materials ", Pisa, 27.09-2.10.1998.
4. J.M.Carpenter and C.A.Pelizzari, Phys.Rew.B 12, 2391 (1975)
5. J.R.D.Copley, Comput.Phys.Commun. 7, 286(1974)
6. J.R.D.Copley, P.Verkerk, A.A.Van Well and H.Fredrikze, Comput. Phys. Commun. 40,337(1986)

STRUCTURAL WORK

Structure of Multi Component Glasses using Diffraction Techniques and Reverse Monte Carlo Modelling

J. Swenson[1], L. Börjesson[1] and R. L. McGreevy[2]

[1] *Department of Applied Physics, Chalmers University of Technology, S-412 96 Göteborg, Sweden*
[2] *Studsvik Neutron Research Laboratory, Uppsala University, S-611 82 Nyköping, Sweden*

Abstract. Structural properties of glasses in general, and complicated multi component glasses in particular, are difficult to extract directly from experimental results. This is partly because global structural investigations by diffraction techniques are only providing a one-dimensional average representation of the structure, usually without any information of partial pair correlations for a single experiment. Furthermore, many glasses are so chemically complicated that it has, so far, been difficult to develop interatomic model potentials between the particles that are accurate enough for reproducing the diffraction data using conventional simulation techniques such as molecular dynamics (MD) and Monte Carlo (MC) simulations. The reverse Monte Carlo (RMC) modelling method overcomes the problem of finding model potentials and instead it makes direct use of the available experimental data. In this way, the RMC method produces three dimensional structural models which are quantitatively consistent with the available data and the applied constraint, e.g. constraints of density, minimum atomic sizes and coordination numbers. As an illustrative example, we show how the RMC method has been used for structural modelling of the fast ion conducting glasses $(AgI)_{0.6}$ $(Ag_2O-2B_2O_3)_{0.4}$ and $(AgI)_{0.75}$ $(Ag_2MoO_4)_{0.25}$. Some advantages and limitations of the method will also be discussed.

1. INTRODUCTION

The structure of glasses is in general not very well known, mainly because of the low degree of structural ordering and the fact that diffraction experiments are only providing one-dimensional average representations of the real three-dimensional structures. For simple glasses, the nearest neighbour distances and coordination numbers have been determined predominantly by neutron or x-ray diffraction. However, even for relatively simple glasses, such as B_2O_3 and SiO_2, the structure beyond the nearest neighbours is less known, and many controversies have continued through the years. They concern, for example, the presence of few membered rings (1,2), the origin of the first sharp diffraction peak in network glasses (3-12), the degree of intermediate range ordering (6,10,12) and similarities to corresponding crystal structures (3,4).

For more complicated multi component glasses the structure is even less known. It is then more difficult to extract structural information directly from diffraction experiments, especially when the use of isotope substitution to obtain elemental contrast is not possible. Instead, the knowledge originates often from atom or molecule specific probes such as nuclear magnetic resonance (NMR), extended x-ray absorption fine structure (EXAFS) and vibrational spectroscopy, which provide local information on a specific part of the structure. However, to get a relevant view of the total structure it is

CP479 *Neutrons and Numerical Methods — N_2M*
edited by M. R. Johnson, G. J. Kearley, and H. G. Büttner
© 1999 The American Institute of Physics 1-56396-838-X/99/$15.00

evident that some kind of structural modelling is needed for most multi component glasses. This can be done by conventional simulation techniques like molecular dynamics (MD) and ordinary Monte Carlo (MC). The problem with these simulation techniques is, however, that they require interatomic model potentials between the particles, which are difficult to define for chemically complicated glasses, which often leads to problems in reproducing the experimental data quantitatively. To overcome this problem and make direct use of the available experimental diffraction data we have taken another approach and used the reverse Monte Carlo (RMC) modelling method, developed by McGreevy and Pusztai 1988 (13,14). In this paper we show how the RMC method can be used for structural studies of fast ion conducting glasses, using the $(AgI)_{0.6} (Ag_2O-2B_2O_3)_{0.4}$ and $(AgI)_{0.75} (Ag_2MoO_4)_{0.25}$ glasses as illustrative examples.

2. THE REVERSE MONTE CARLO METHOD

The difference between RMC and ordinary Monte Carlo (MC) simulations is that RMC use experimental data instead of interatomic potential in the minimizing procedure. RMC uses a standard Metropolis Monte Carlo (MMC) algorithm (15) (Markov chain, periodic boundary conditions etc.) but with the "sum of squares" difference between the measured structure factors and those calculated from the RMC configuration as a "driving parameter" in place of the energy (see Refs. 16 and 17 for more details of the method). Data from different sources (neutron, x-rays, EXAFS) may be combined. In this way, the RMC method produces three dimensional models of the structure of disordered materials that agree quantitatively with the available diffraction data and the measured density, providing that the data do not contain significant systematic errors. The fact that the method does not require any interatomic potential gives a great advantage compared to conventional simulation techniques for structural modelling of materials, where the existing interatomic model potentials cannot provide quantitative agreement with the experimental data.

2.1 The Use of Constraints

Before one begins the actual RMC simulation it is convenient to run a hard sphere Monte Carlo (HSMC) simulation to create a computer configuration which is consistent with the applied contraints. These constraints can be such as closest approach distance of two particles, maximum bond length, bond angles and number of nearest neighbours. The constraints are usually determined from the experimental diffraction data or other experimental techniques such as NMR and Raman spectroscopy. Thus, for the present glasses the constraints serve to produce initial configurations that are consistent with previous knowledge of the glasses, i.e. a 3-dimensional network with 3 and 4-coordinated B in the case of the $(AgI)_{0.6} (Ag_2O-2B_2O_3)_{0.4}$ glass and the presence of molecular MoO_4^{2-} ions for the $(AgI)_{0.75} (Ag_2MoO_4)_{0.25}$ glass. In the case of the network connectivity of the silver borate glass we have used magic angle spinning NMR data (18,19), which appear to show that for all undoped and salt doped diborate compositions approximately 45% of the borons are four-coordinated and 55% are three-coordinated to oxygen, while almost all oxygens are bridging between two boron atoms. For the silver molybdate glass we started with a random distribution of tetrahedral MoO_4^{2-} ions. Both glasses contained about 4000 atoms and the closest atom-atom distances (of all the ten partial pair correlations) were taken from experimental radial distribution functions and chemical knowledge (see Refs. 20 and 21). The Ag^+ and I^-

ions were randomly added into the computer box and after the applied constraints had been fulfilled by the HSMC simulation the actual RMC simulation could begin.

3. RESULTS AND DISCUSSION

In Fig. 1 we compare the experimental and RMC produced neutron and x-ray structure factors for the $(AgI)_{0.6}$ $(Ag_2O-2B_2O_3)_{0.4}$ and $(AgI)_{0.75}$ $(Ag_2MoO_4)_{0.25}$ glasses (20,21). Since the overall structure factors are well reproduced and the atomic scattering coefficients are very different (i.e. the different atomic pair correlations are very differently weighted) in the neutron and x-ray diffraction experiments, the RMC configurations should contain the essential structural features of the investigated glasses. The small differences between the experimental and RMC produced neutron structure factors are due to some systematic errors in the neutron data.

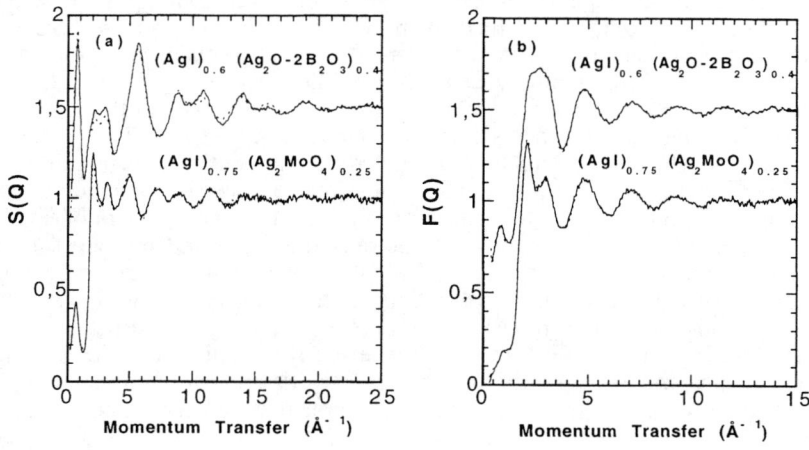

FIGURE 1. (a) Neutron and (b) x-ray structure factors of $(AgI)_{0.6}$ $(Ag_2O-2B_2O_3)_{0.4}$ (shifted by 0.5 for clarity) and $(AgI)_{0.75}$ $(Ag_2MoO_4)_{0.25}$.

From the structural models, it is in principle possible to extract a lot of structural information, e.g. all partial pair correlation functions (10 for each of the glasses presented here) and various bond angle distributions. However, one should be careful and realize that the atomic configurations are models of the real structure and that they do not contain more information than put in with the experimental data and the constraints. Without the use of constraints the RMC produced models are only providing the information present in the experimental data, which is relatively limited for a single data set. It is, in fact, unlikely that the RMC model would even produce a physically sensible structure in that way. Therefore, it is very important to use additional constraints in order to reduce the number of possibilities to fit the data. At least constraints for closest atom-atom distances have to be applied. With correctly used constraints it is possible to exclude unphysical structures and to come close to a "real" typical structure. The problem is, however, that one rarely has sufficient structural and chemical knowledge to provide detailed constraints. One can partly overcome this problem by using several experimental data sets with different atomic scattering factors (i.e. using isotope substitution, both neutron and x-ray data, differential anomalous x-ray scattering etc.), since additional structural data sets reduces the number of possible configurations. Thus, for best results one should use all the available experimental data and as many additional constraints (which has to be correct) as possible.

With the additional constraints and the two data sets used for the glasses of $(AgI)_{0.6}$ $(Ag_2O\text{-}2B_2O_3)_{0.4}$ and $(AgI)_{0.75}$ $(Ag_2MoO_4)_{0.25}$, the ten partial structure factors (see Figs. 2(a) and 2(b)) and pair correlation functions should reflect the main features of the real structures. In the present cases, different initial structures and slightly different or additional constraints did not cause any major change of the partial structure factors and pair correlation functions. Therefore, we are, for example, able to determine (20,21) which partial structure factors that mainly contribute to the much debated first (sharp) diffraction peak (FSDP) located at the low Q-values 0.8 Å^{-1} and 0.65 Å^{-1} for the borate and moybdate glasses, respectively (see Fig. 1).

For the AgI doped borate (20) and phosphate (22) glasses it was evident (see Fig. 2(a)) in the case of the borate glass) that the FSDP occurs predominantly in those partials related to the glass network, e.g., B-B, B-O and O-O. In these glasses the FSDP arises because the phosphate chains or diborate segments, which remain covalently bonded, are pushed further apart by the introduced salt ions. Due to the large covalent character of the silver ions it seems that the silver and iodine ions participate in the cross-linking between neighbouring chain segments, forming local B-O-Ag-I-Ag-O-B ordering with a mean distance between two neighbouring network segments corresponding to the position of the FSDP, i.e. about 8-9 Å. Thus, most of the silver ions are coordinated to both oxygens and iodine ions and there is no presence of any AgI clusters of significant sizes (> 10 Å), which has been proposed to be related to the FSDP in the neutron data (see for example Ref. 23). In fact, this can be qualitatively concluded directly from the experimental x-ray data, since any explanation of the FSDP in terms of AgI clusters would have given rise to an enormously intense prepeak in the x-ray structure factor. In contrast, the FSDP in the x-ray F(Q) is actually very tiny (see Fig. 1(b)). Due to the salt induced expansion and ordering of the glass network the structure shows clearly pronounced microscopic pathways for ion conduction between the chain-like network structure.

In the case of the molybdate glass the origin of the FSDP is slightly different since there is no FSDP in the Mo-Mo partial structure factor (see Fig. 2(b)). The reason for this is that the MoO_4^{2-} are almost randomly distributed (in contrast to the PO_4 or BO_3 and BO_4^- units in the phosphate and borate glasses) but have some orientational correlations (producing the FSDP in the Mo-O and O-O for example) which must be caused by the cross-liking of Ag. This explains also the rather narrow glass forming region for the molybdate glass, because if the AgI content is too low then too many MoO_4^{2-} will be close together and the orientational correlations will lead to crystallisation. On the other hand, if the AgI concentration is too high then there will not be enough (or too weak) cross-linking and the network will not be inter-connected, also leading to crystallisation.

Concerning the short range order the RMC were able to obtain reasonable values for most of the interatomic coordination numbers for both the borate and molybdate glasses. If the coordination numbers can be determined directly from the experimental data the RMC produced model always reproduce these values (within the experimental errors). The RMC model will, however, also contain much additional information, e.g. reasonable determinations of coordination numbers which are not possible to extract directly from the experimental data. The reason for this is that all the partial pair correlation functions must be consistent with one another, which means that some information on partials can be obtained for underdetermined cases.

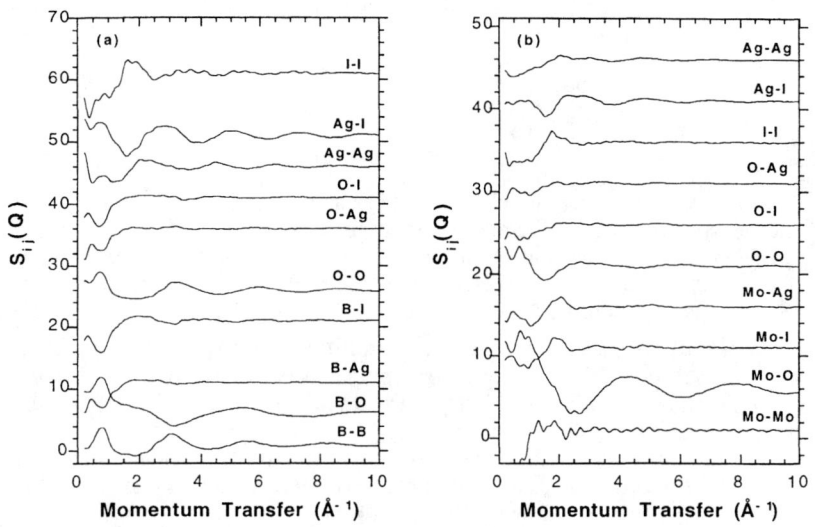

FIGURE 2. Partial structure factors, $S_{ij}(Q)$, determined from RMC models for (a) $(AgI)_{0.6}$ $(Ag_2O\text{-}2B_2O_3)_{0.4}$ and (b) $(AgI)_{0.75}$ $(Ag_2MoO_4)_{0.25}$. Upper curves have bee shifted vertically for clarity.

One limitation of the RMC method is that it tends to produce the most disordered structure that is consistent with the data and constraints, i.e. the configurational entropy is maximised. This means, for example, that the distributions of bond lengths (or nearest neighbour distances), bond angles (calculated from nearest and second nearest distances) and coordination numbers tend to be more distributed in the model than in the "real" glass. The distributions of bond lengths and bond angles can in many cases be determined with relatively good accuracy directly from the experimental data and in those cases one can use rather narrow constraints to narrow the distributions. It is, however, not possible to determine the distribution of coordination numbers from the diffraction data since the data provides only an average value, instead it is necessary to use constraints based on other experimental results or chemical knowledge. For covalent network glasses such coordination number constraints are usually quite obvious (for example the constraints we used for the boron-oxygen network of the $(AgI)_{0.6}$ $(Ag_2O\text{-}2B_2O_3)_{0.4}$ glass), but it is more difficult to determine the "bonding" constraints for ionic glasses or the salt ions in the present glasses.

In this paper we introduce a new kind of constraint which is appropriate to use for interatomic correlations involving ions. The constraint is based on bond valence sums calculated from pseudopotentials. The valence sum

$$V = \sum_X s_{Ag-X} \qquad (1)$$

for each Ag$^+$ ion is calculated from empirical bond-length bond-valence equations. We have chosen the parameter set by Radaev, Fink and Trömel (24,25) given in Eq. 2.

$$s_{Ag-O} = \exp\left[\frac{1.89 \text{Å} - R_{Ag-O}}{0.33 \text{Å}}\right]$$

$$s_{Ag-I} = \exp\left[\frac{2.08 \text{Å} - R_{Ag-I}}{0.53 \text{Å}}\right] \quad (2)$$

The authors obtained the parameter values by calculating the Ag$^+$ valence sums for different crystal structures and fitting the average values to 1, which is the expected valence sum for a mono valent cation such as Ag$^+$. All oxygens and iodine ions up to a distance of 6 Å ($R_{Ag-X} < 6$ Å) from the respective Ag site are included. For a physically correct structure it is expected that the valence sum V should be relatively close to a target valence of 1. Thus, by using a constraint which minimises the valence difference $\Delta V = |V-1|$ for each Ag$^+$ ion it is possible to minimize unphysical Ag coordinations. However, one should note that it has not been possible to fit the diffraction data with a strict constraint on the valence sum, indicating that the individual Ag$^+$ ions in the real glasses show some deviations from the empirical valence sum of 1. Generally, the valence sum tends to be slightly lower than 1 and we have not been able to observe values significantly higher than 1 for more than a few Ag$^+$ ions. For the RMC produced configurations where we did not include the bond valence sum constraint the distributions of bond valences were slightly broader, with a few more Ag$^+$ ions having unphysically low and high bond valences.

4. CONCLUSION

In this paper we have shown how the structure of multi component glasses, such as the fast ion conducting glasses $(AgI)_{0.6} (Ag_2O-2B_2O_3)_{0.4}$ and $(AgI)_{0.75} (Ag_2MoO_4)_{0.25}$, can be modelled using diffraction data and the RMC method. Since RMC does not require any interatomic model potential (which obviously has both advantages and disadvantages) the method is most useful for materials where the existing interatomic model potentials cannot provide quantitative agreement with the experimental data. For the present glasses we were able to determine the origin of the FSDP and find structural properties related to the ionic conductivity. For the borate glass we found that the introduced salt ions cause an expansion and ordering of the B-O network, where clearly pronounced pathways for ion conduction are shown between the borate segments.

The RMC method is not producing any unique or "correct" structure, which in fact no technique will do. It is only producing a model which is quantitatively consistent with the available data and the constraints applied. However, with several experimental data sets and additional constraints the number of possibilities to fit the data will be substantially reduced, and the final model has to contain the main structural features of the investigated material. Finally, we have introduced a new kind of constraint which is appropriate to use for ionic glasses and interatomic correlations involving ions. The constraint, which is based on bond valence sums calculated from pseudopotentials, ensures that the bond valences of the ions will be reasonable.

ACKNOWLEDGEMENTS

This work was financially supported by the Swedish Natural Science Research Council.

REFERENCES

(1) Johnson, P.A., Wright, A.C., Sinclair, R.N., *J. Non-Cryst. Solids* **50**, 281 (1982).
(2) Swenson, J., and Börjesson, L., *Phys. Rev. B* **55**, 11138 (1997).
(3) Busse, L.E., *Phys. Rev B* **29**, 3639 (1984).
(4) Gaskell, P.H., and Wallis, D.J., *Phys. Rev. Lett.* **76**, 66 (1996).
(5) David, M.F., and Leadbetter, A.J., *Phil. Mag. B* **44**, 509 (1981).
(6) A. Uhlherr, A., and Elliott, S.R., *J. Phys.: Condens. Matter* **6**, L99 (1994); *Phil. Mag. B* **71**, 611 (1995).
(7) P. S. Salmon, P.S., *Proc. R. Soc. Lond.* **A445**, 351 (1994).
(8) Price, D.L., Moss, S.C., Reijers, R., Saboungi, M-L., Susman, S., *J. Phys.: Condens. Matter* **1**, 1005 (1989).
(9) Elliott, S.R., *Phys. Rev. Lett.* **67**, 711 (1991); *J. Phys.: Condensed Matter* **4**, 7661 (1992).
(10) Swenson, J., and Börjesson, L., *J. Non-Cryst. Solids* **223**, 223 (1998).
(11) Iyetomi, H., and Vashishta, P., *Phys Rev. B* **47**, 3063 (1993).
(12) Nakano, A., R. K. Kalia, R.K., Vashishta, P., *J. Non-Cryst. Solids* **171**, 157 (1994).
(13) McGreevy, R.L., and Pusztai, L., *Mol. Simul.* **1**, 359 (1988).
(14) Keen, D.A., and McGreevy, R.L., *Nature* **344**, 423 (1990).
(15) Metropolis, N., Rosenbluth, A.W., Rosenbluth, M.N., Teller, A.H., Teller, E., *J. Phys. Chem.* **21**, 1087 (1953).
(16) McGreevy, R.L., Ann. Rev. Mat. Sci. 22, 217 (1992).
(17) McGreevy, R.L., Nucl. Inst. Meth. in Phys. Res. A 354, 1 (1995).
(18) Chiodelli, C., Magistris, A., Villa, M., Bjorkstam, J.L., *J. Non-Cryst. Solids* **51**, 143 (1982).
(19) Feller, S.A., Dell, W.J., Bray, P.J., *J. Non-Cryst. Solids* **51**, 21 (1982).
(20) Swenson, J., Börjesson, L., McGreevy, R.L., Howells, W.S., *Phys. Rev. B* **55**, 11236 (1997).
(21) Swenson, J., McGreevy, R.L., Börjesson, L., Wicks, J.D., Howells, W.S., *J. Phys.: Condens. Matter* **8**, 3545 (1996).
(22) Wicks, J.D., L. Börjesson, McGreevy, R.L., Howells, W.S., Bushnell-Wye, G., *Phys. Rev. Lett.* **74**, 726 (1995).
(23) Rousselot, C., Malugani, M.P., Mercier, R., Tachez, M., Chieux, P., Pappin, A.J., Ingram, M.D., *Solid State Ionics* **78**, 211 (1995).
(24) Radaev, S.F., Fink, L., Trömel, M., *Z. Kristallogr. Suppl.* **8**, 628 (1994).
(25) Adams, S., Maier, J. *Solid State Ionics* **105**, 67 (1998).

Microscopic Structure of Amorphous Carbon. Tight-Binding Molecular Dynamics Study

S. Kugler, K. Koháry and I. László

Department of Theoretical Physics, Institute of Physics, Technical University of Budapest, H-1521 Budapest, Hungary, E-mail: kugler@phy.bme.hu

Abstract. In our computer simulation the preparation process was similar to the atom-by-atom deposition onto a substrate which is a standard procedure to prepare amorphous carbon. For the calculation of the interactions between carbon atoms a tight-binding potential was used. All parameters and functions of the potential were fitted to the results of local density functional calculations for graphite, diamond and linear chains. The Newton equations of motion were solved by the Verlet algorithm. The amorphous structures with fivefold, sixfold and sevenfold rings and chain-like atomic arrangement were obtained on diamond (111) surface.

INTRODUCTION

Amorphous semiconductors attract continuing attention because of their importance in technological applications. Among these materials, the amorphous form of carbon (a-C) holds a special and unique position. The ability of carbon atoms to have sp^3, sp^2 and sp bonding configurations leads to a large variation in structure and in electronic properties, and furthermore, produces a wide range for macroscopic parameters of a-C film. Several electron and neutron diffraction measurements were employed on a-C samples in order to determine the atomic arrangement. The first well-know experiment was carried out on an evaporated sample using electron diffraction by a Japanese group [1]. On the basis of these measured data the authors proposed a micro-crystalline model for amorphous carbon structure, consisting of graphitic and diamond-like regions. The first neutron diffraction measurement was carried out by Mildner and Carpenter [2]. They estimated a 5 percent diamond-like atomic configuration in their sample. F. Li and J.S. Lannin published also a neutron diffraction measurement on a sputtered a-C sample [3]. The radial distribution functions (RDF) obtained differed in detail from the theoretical models. Another diffraction study of amorphous carbon material prepared using a filtered vacuum-arc method proposed near one-hundred percent of sp^3

atomic configuration [4]. It was measured at D4 spectrometer at the ILL, Grenoble. Neutron diffraction was employed on an anomalous a-C sample [5] in order to determine the atomic arrangement and to decide whether the structure is temperature dependent [6]. The sample was prepared by arc-evaporation of graphite rod under a pressure of 10^{-6} atm at the Gifu University.

STRUCTURAL MODELING

There are two main possibilities for structural modeling. The first one is the Monte Carlo (MC) method. Metropolis et al. were the pioneers of this kind of computer simulation [7]. Recently a new version of this method - the so-called Reverse Monte Carlo simulation - has been developed by McGreevy and Pusztai [8], which is convenient for the investigation of amorphous materials. It is based on the results of diffraction measurements. The algorithm is very similar to the standard Metropolis Monte Carlo technique [7]. This method was applied for constructing large a-Si models [9]. Second, Molecular Dynamics (MD) is another useful tool for describing non-crystalline systems. To integrate the MD equations of motion the Verlet algorithm [10] is usually used with a time step equal to about a picosecond. By MD the dynamics of the systems can be followed, therefore it is useful to model fundamental phenomena like growth from vapor phase to a surface.

We consider two structural models of amorphous carbon. Galli and her co-workers constructed an amorphous carbon structure by computer simulation [11]. They carried out a first principle Molecular Dynamics simulation for 54 atoms using 2.0 g/cm^3 of macroscopic density. The inter-atomic potential was constructed from the electronic ground state treated with density-functional techniques. The model was obtained by rapid cooling of liquid carbon. As similar method was used by Stephan and his co-workers [12] in order to generate a-C models having different densities in the range of 2.0-3.52 g/cm^3. They applied a semi-empirical density-functional approach. These larger models contained 128 carbon atoms.

METHOD

We simulated the growth procedure of the amorphous carbon structure, on a diamond (111) surface. Our motivation was the following: amorphous carbon usually is grown by vapor-growth technique and it is not so easy to prepare from the liquid phase using rapid quenching. The formation of amorphous carbon films on a substrate was simulated similarly to atom-by-atom deposition. In MC and MD cases the crucial point is the applied local potential. The MD simulation was performed with the tight-binding Hamiltonian of Xu, Wang, Chan and Ho [13] for carbon systems. All the parameters and functions of this method were fitted to the results of local density functional calculations for graphite, diamond and linear chains. Periodic boundary conditions in x and y directions were adopted to avoid surface effects.

Same potential was applied for the formation of cage-like fullerenes in helium atmosphere using tight-binding MD simulation [14]. The He-He and C-He interactions were described by pairwise 6-12 Lennard-Jones potentials.

In our computer experiment there were three different sets of carbon atoms. The first one was the atomic current of carbon atoms forwarded to the target of a (111) diamond surface. The second set was the upper part of the substrate, which contained atoms at 0 K of initial temperature. The atoms of the third set formed the bottom part of the substrate. These atoms were fixed at their ideal lattice sites.

During the growth procedure a carbon atom started to move forward to the

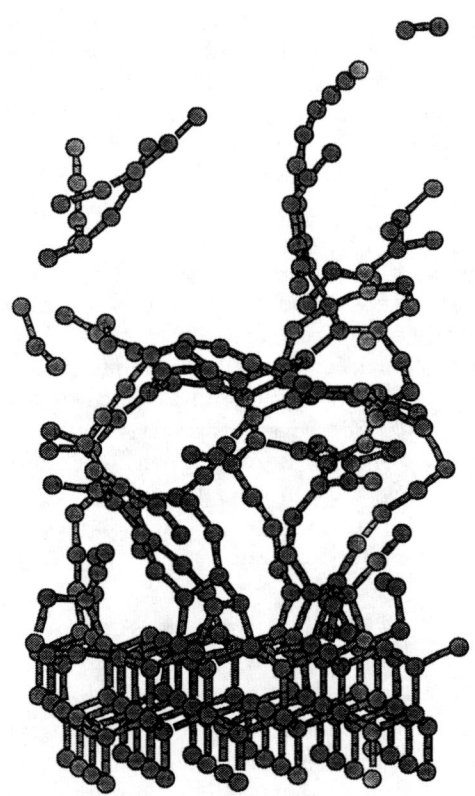

FIGURE 1. The final structure in the unit cell of the atom-by-atom deposition. The substrate at 0 K is at the bottom of the Figure. The kinetic energy of the incoming atoms was relative low (6.5 eV). This model yields a chain-like amorphous structure. It contains 286 carbon atoms including 120 substrate atoms.

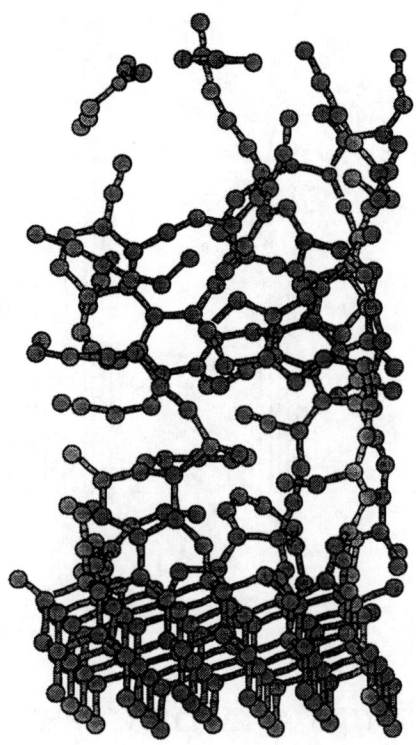

FIGURE 2. The second structure. The average kinetic energy of incoming atoms were 13 eV. It contains 323 carbon atoms.

target surface. Its initial x, y coordinates and the initial velocities were randomly distributed. As the cut-off distance of our tight-binding potential was 2.6 Å, to speed up the simulation, the initial z coordinates in the current were placed at 3.6 Å over the target. When the particle arrived within the critical distance at the substrate, which was a little bit greater than the bond distance, the particle became part of the bonded carbons. To simulate inelastic collision by the target surface the velocity of the current atoms were decreased in each MD step by a given factor which was a little bit lower than 1. It was a simple model of energy dissipation on the substrate surface.

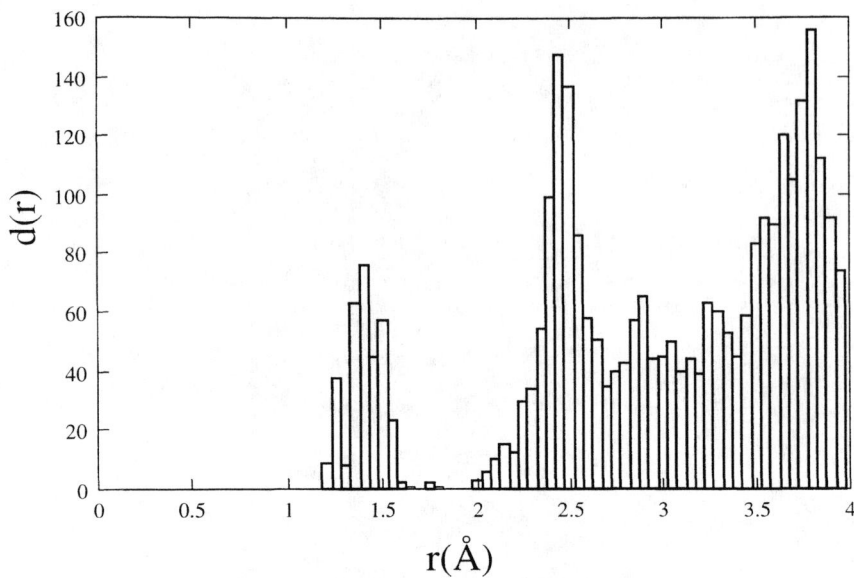

FIGURE 3. The radial distribution function for the model where the kinetic energy was 13 eV.

RESULTS AND DISCUSSIONS

The results of a-C film growth simulation is presented for two different cases. The total simulation time of the atom-by-atom deposition was usually 5-10 psec and the time step was equal to 0.5 femtosecond. Figure 1 shows the final structure of the first simulation in the unit cell. The random current came from the z direction. The kinetic energy of incoming atoms was relative low (6.5 eV). Figure 2 displays a model where the kinetic energy was 13 eV. This energy is close to the value used during the a-C sample preparation. In both cases the substrate consist of 120 carbon atoms. The structure of the first model was a chain-like amorphous structure while the second one contains more sp^2 and sp^3 atomic sites and rings. It means that the higher incoming energy produces less chains and more rings. Similar experimental results have been recently published [15,16]. In Figure 3 we present the calculated radial distribution function d(r). RDF exhibit quantitative agreement with experimental results measured by neutron diffractions mentioned above within the interval of 1-4 Å.

In summary we have presented the formation of amorphous carbon structures under various simulation conditions. The final atomic arrengements have fivefold, sixfold and sevenfold membered rings and chain-like structure depending on the kinetic energy of incoming atoms.

ACKNOWLEDGEMENT

This work has been supported by the Fund OTKA (Grant No. T024138, T025017).

REFERENCES

1. J. Kakinoki, K. Katada, T. Hanawa et al., *Acta Cryst.* **13**, 171 (1960).
2. D.F.R. Mildner and J.M. Carpenter *J. of Non-Cryst. Solids* **47**, 391 (1982).
3. F. Li. and J.S. Lannin, *Phys. Rev. Lett.* **1905**, 65 (1990).
4. P.H. Gaskell, A. Saeed, P. Chieux, and D. R. McKenzie, *Phys. Rev. Lett.* **1286**, 67 (1991).
5. K. Shimakawa, K. Hayashi, T. Kameyama, T. Watanabe and K. Morigaki, *Phil. Mag. Lett.* **64**, 375 (1991).
6. S. Kugler, K Shimakawa, T. Watanabe, K. Hayashi, I. László, and R. Bellissent, *J. of Non-Cryst. Solids* **164-166**, 831 (1993).
7. N. Metropolis, A.W. Rosenbluth, M.N. Rosenbluth, et al., *J. Chem. Phys.* **21**, 1087 (1953).
8. R.L. McGreevy and L. Pusztai, *Molec. Sim.* **1**, 369 (1988).
9. S. Kugler, L. Pusztai, L. Rosta, P. Chieux and R. Bellisent, *Phys. Rev. B* **48**, 7685 (1993).
10. L. Verlet, *Phys. Rev.* **159**, 98 (1967).
11. G. Galli, R.M. Martin, R. Car and M. Parrinello, *Phys. Rev. Lett.* **62**, 555. (1989).
12. U. Stephan, Th. Frauenheim, P. Blaudeck et al., *Phys. Rev. B* **49**, 1489 (1994).
13. C.H. Xu, C.Z. Wang, C.T. Chan, K.M. Ho, *J. Phys. C* **4**, 6047 (1992).
14. I. László, *Europhys. Lett.* **44**, 741 (1998).
15. I. Pócsik, M. Koós, M. Hundhausen and L. Ley, "Excitation Energy Dependent Raman and Photoluminescence Spectra of Hydrogenated Amorphous Carbon" in: *Amorphous Carbon: State of Art*, Ed. by S.R.S. Silva, J. Robertson, G.A.J. Amaratunga and W. I. Milne, Word Scientific, Singapore, 1998, pp. 224-233.
16. I. Pócsik, M. Koós, O. Berkesi and M. Hundhausen," Advantage of Infrared Excitation in Raman Spectroscopy of Hydrogenated Amorphous Carbon Films" in: *Proceedings of the Sixteenth International Conference on Raman Spectroscopy Edited by A.M. Heyns*, John Wiley and Sons, New York, 1998, Suppl. Vol. pp.: 66-67.

Monte-Carlo sorption and neutron diffraction study of the filling isotherm in clathrate hydrates

Alice Klapproth[*], Bertrand Chazallon[*,#] and Werner F. Kuhs[*]

[*]*Mineralogisch-Kristallographisches Institut, Universität Göttingen, Goldschmidtstr.1, 37077 Göttingen, Germany*
[#]*Laboratoire de Physico-Chimie des Matériaux Luminescents, Université Claude Bernard-Lyon I, 43 boulevard du 11 Novembre 1918, 69622 Villeurbanne Cédex, France*

Abstract. We are interested in the thermodynamics of the gas filling of clathrate hydrates. In order to determine the pressure-dependent filling of the cages, neutron powder diffraction experiments on N_2 and CO_2 clathrates were performed. Interaction potentials were refined by comparing the experimentally determined fillings with those generated by MC-sorption calculations. Unsatisfactory agreement between experiment and simulation is observed when using the widely employed SPC water-water interaction potential.

Introduction

Clathrate hydrates are inclusion compounds in which the host structure constitutes a framework of small and large cages made up of H-bonded water molecules filled by a variety of gas species. The empty clathrate hydrate framework is stabilised at elevated pressures and low temperatures by full or partial gas occupancy. The fugacity-dependent filling of the hydrate structure is generally assumed to follow a Langmuir isotherm as described by van der Waals and Platteeuw (1). The possibility to calculate filling and thermodynamic stability of clathrate hydrates is of great technological interest: clathrate formation can lead to blockages in natural gas pipelines but is also a welcome energy resource in the form of methane hydrate deposits in permafrost regions and in the deep sea floor. It is therefore important to check the validity of this generally accepted but unproven theory. We have started a neutron diffraction programme to establish for the first time the experimental filling as a function of gas fugacity for a variety of gases. In the case of CO_2, we have strong evidence that the filling deviates from a Langmuir behaviour (2). Clathrates of mixed gases cannot be investigated easily by diffraction methods due to the time-space averaging; in this case relative numbers of occupation and gas fractionation are gained from Raman spectroscopy. The best access to mixed clathrates seems to be through MC-simulations of the filling. However, before co-enclathration processes can be modelled, the pure systems have to be understood and the appropriate molecular interaction potentials established. Once the pure system can be modelled in a satisfactory way, the corresponding interaction potentials may be used to simulate gas adsorption processes in mixed clathrates or on ice surfaces, the latter being of considerable importance for improving the gas analysis of deep ice cores (3,4). In this paper, we compare our experimentally established cage filling for CO_2 and N_2 clathrate hydrates, which represent type I (SG: $Pm\overline{3}n$) and type II (SG: $Fd\overline{3}m$) structures in the von Stackelberg classification (5) respectively, with that obtained from MC-sorption simulations using widely accepted interactions potentials.

Experimental

Synthetically obtained N_2 and CO_2 clathrates have been analysed *in-situ* on the neutron powder diffractometer D2B at ILL/Grenoble at a wavelength of 1.59 Å. The clathrates were prepared with D_2O usually at a temperature of 273K using pressure cells operating in a pressure regime up to 3000 bar. The resulting diffraction patterns were treated in Rietveld refinements using GSAS (6) to determine the detailed structure and the site occupancy of the gas molecules in the different cages. The final weighted profile R-factors are in the vicinity of 0.02. Molecular dynamics simulations with subsequent energy minimizations were performed to establish the positions of the disordered gas molecules within the host lattice (7). Instead of the usually assumed filling of only one gas molecule per large cage, we have found clear evidence for a partial double occupancy in the case of N_2 clathrates (8). A corresponding two-constant model for single and double occupancy gives a reasonable fit to the data (Fig.1). The filling of the large cages for the CO_2 clathrate nearly follows that of a Langmuir isotherm (2,9). However, as pressure increases the small cages are far from approaching a complete filling, thus violating the generally accepted solid-solution theory of van der Waals and Platteeuw (Fig.2).

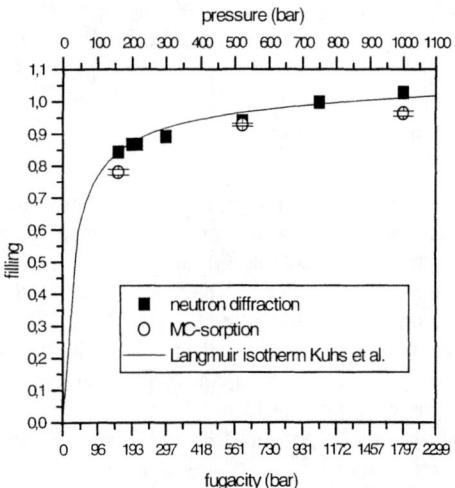
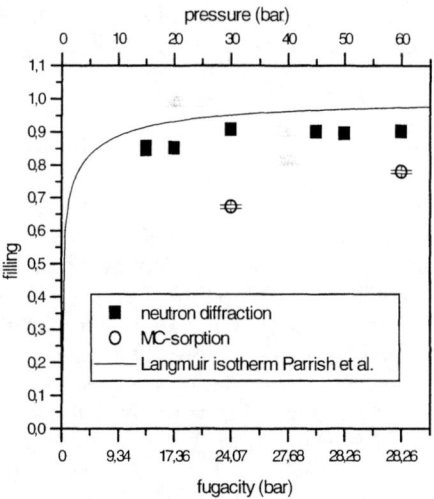

FIGURE 1. Comparison between observed and calculated cage filling of N_2 clathrates (Type II) at 273K. The N_2 VDW radius was optimised to 4.1 Å.

FIGURE 2. Comparison between observed and calculated cage filling of CO_2 clathrates (Type I) at 273K.

Simulation Environment

Simulations were performed on a Silicon Graphics workstation Indigo2 using Cerius2. The MC-sorption routine implements a rapid Monte-Carlo (MC) statistical mechanics method and simulates adsorption of small molecules (gases) into porous 3D crystal structures such as clathrate hydrates. We used the "fixed pressure–variable loading" method which is based on a grand canonical ensemble. The initial configuration contains no sorbate molecule and each subsequent configuration is generated by either a translation, rotation, creation or destruction of sorbates. The energy calculation is confined to the intermolecular energy, as both the framework and the sorbate are held internally rigid during the simulation.

The hydrate structure is simulated using the SPC-model of water (10), while for N_2 and CO_2 the models published as X1 (11) and 5-charge-model (12) were respectively employed. The simulation was performed using one or eight cubic unit cells with periodic boundary conditions, with the water molecules arranged according to the ice rules. A single cell for N_2 clathrate comprised 136 water monomers whilst a single cell for CO_2 comprised only 46. The simulations were carried out at a temperature of 273K and fixed pressures of 160, 523 and 1000 bar for the N_2 clathrate and at 30 and 60 bar for CO_2 clathrate. The unit cell constants were adjusted to the pressure according to values obtained from experimental data. Spherical cut-offs at 8.0 Å were found sufficient and were subsequently used in evaluating the sorption-energy, considering both van der Waals and Coulombic (dipole-quadrupole) interactions. The sorbate-framework electrostatics were evaluated using the Ewald summation method. Configurations with bad contacts, sorbate and framework atoms closer to each other than half of their van der Waals radius, were rejected from the energy calculation. The MC computations were performed for over 5 million moves with the first million dropped in the averaging processes. The step size was approximately 0.5 Å ($\approx$ one-tenth of cavity radius) for translation and 180° for rotation with a rescale factor of 1.05, a success rate of 50 % and a rescale frequency of one per 10000 configurations.

Results and Discussion

The experimental gas fillings from *in-situ* neutron powder diffraction were established with a precision of 0.5 to 1 % and an accuracy of 1-2 % (2) and thus provide an excellent test for the validity of the simulations. Initial simulations performed using models commonly cited in literature (VDW radii: 3.318 Å for N, 3.17 Å for C and 3.40 Å for O in the case of CO_2) gave strong disagreement with experimental numbers. The calculated cage fillings of CO_2 clathrate were always lower than those obtained from neutron diffraction (Fig.2). In the case of N_2 clathrate, initial runs at a pressure of 160 bar gave double occupancy of the large cage as opposed to the experimentally determined single occupancy at this pressure. N_2 clathrate hydrate is an especially good case for testing the pressure dependent filling of the cages due to the possibility of double occupancy. Better agreement was achieved for the 500 bar data by increasing the van der Waals radius of N_2 from 3.318 Å to 4.1 Å (Fig.1). Unfortunately, the simulated fillings at the other pressures, using the enlarged N_2 radius, were in poor agreement with the experimental data. At a pressure of 160 bar the modified VDW radius appeared to be too small whilst at 1000 bar it appeared to be too large, as evidenced by the failure to produce double occupancy for the large cages. These systematic deviations could be alleviated only by changing the van der Waals radii in the SPC water.

The interaction energies for N_2 and CO_2 with the host lattice ("sorption energy") have been calculated on the basis of the Lennard-Jones-Devonshire model to 4.01 kcal/mol and 7.08 kcal/mol (13), which gives an energy difference of approximately 3 kcal/mol. The energies may also be estimated from thermodynamic data of the heat of evaporation of the constituents, gas and water (14). Using Troutons formula to estimate the heat of vaporisation for N_2 and CO_2, one obtains the difference in the sorption energy between the corresponding hydrates of approximately 2 kcal/mol. It is noteworthy that the sorption energies found in our MC-simulations for N_2 (2.9 and 3.6 kcal/mol for large and small cages) and CO_2 (4.2 kcal/mol at maximum; small and large cages not separated) are somewhat lower than the previously calculated values and the difference between the two is also smaller. This discrepancy may be partly due to unoptimised interaction potentials used in the MC-calculations. In addition, it is experimentally established by our neutron work that the relatively large CO_2 molecule does not fit very well into the small cage (2). Our MC simulations yield sorption energies of similar magnitude for CO_2 in the small and large cages, while a difference of approximately 0.7 kcal/mol is found for N_2 which in turn leads to the observed rather small difference in sorption energy between N_2 and CO_2. Further work is in progress to fine tune the interaction parameters used in these and related systems such as the noble gas (Ar, Kr, Xe) clathrate hydrates.

References

1. van der Waals, J. H., and Platteeuw, J. C., *Adv. Chem. Phys.* **2**, 1 (1959).
2. Kuhs, W. F., Chazallon, B., Klapproth, A., and Pauer, F., *Rev. High Pressure Sci. Technol.* **7**, 1147-1149 (1998).
3. Anklin, M., Schwander, B., Tschumi, J., and Fuchs, A., *J. Geophys. Res.* **102**, C12, 26,539-26,545 (1997).
4. Kuhs, W. F., Klapproth, A., and Chazallon, B., "Chemical physics of air clathrate hydrates in ice cores" in Proceedings of the International Symposium on Physics of Ice-Core Recordes, Shikotsukohan, September 1998, Japan (1999).
5. v. Stackelberg, M., and Mueller, H. R., *Z. Elektrochem.* **58**, 25 (1954).
6. Larsen, A., C., and Von Dreele, B., "GSAS-General Structure Analysis System", University of California, 1985-1995.
7. Chazallon, B., Klapproth, A., and Kuhs, W. F., "Molecular-dynamics modelling and neutron diffraction study of the site disorder in air clathrate hydrate" in Proceedings of the Neutrons & Numerical Methods, Grenoble, December 1998, France (in press).
8. Kuhs, W. F., Chazallon, B., Radaelli, P. G., and Pauer, F., *J. Inclus. Phenom.* **29**, 65 (1997).
9. Parrish, W., R., and Prausnitz J., M., *Ind. Eng. Chem.* **11** (1), 26-34 (1972).
10. Berendsen, H. J. C., Postma, J. P. M., van Gunsteren, W. F., and Hermans, J., *Intermolecular Forces* by D. Riedel Publishing Company 1981, 331-342.
11. Murthy, C. S., Singer, K., Klein, M. L., and Mc Donald, I. R., *Molecular Physics* **41**, 6, 1387-1399 (1980).
12. Murthy, C. S., O'Shea, S. F., and Mc Donald, I. R., *Molecular Physics* **50**, 531-541 (1983).
13. Davidson, D., W., "Water - A Comprehensive Treatise" **3**, edited by Franks, F., Chapter 3, 115-234.
14. v. Stackelberg, M., *Naturwissenschaften* **36**, 12, 359-362 (1949).

Molecular-dynamics modelling and neutron diffraction study of the site disorder in air clathrate hydrates

Bertrand Chazallon[*,#], Alice Klapproth[*] and Werner F. Kuhs[*]

[*]*Mineralogisch-Kristallographisches Institut, Universität Göttingen, Goldschmidtstr.1, 37077 Göttingen, Germany.*
[#]*Laboratoire de Physico-Chimie des Matériaux Luminescents, Université Claude Bernard-Lyon I, 43 boulevard du 11 Novembre 1918, 69622 Villeurbanne Cédex, France.*

Abstract. We present the results of MD-simulation runs with subsequent quenches for clathrate hydrates using SPC water in order to model properly the crystallographic site disorder of the guest molecules in the water cages. A procedure is described to transform the results of the quench (symmetry P1) into the proper space-time averaged space group (Fd$\overline{3}$m) of the clathrate hydrate. The resulting disorder models are compared with the outcome of crystallographic structure refinements (R-factors, Fourier maps) from our neutron powder diffraction data. A correct description of the disorder is important for a reliable determination of the pressure-dependent cage filling.

Introduction

The analysis of gas concentrations imprisoned in ice cores from the (ant)arctic ice sheets provides a way to reconstruct the paleoatmosphere and its changes. However, due to the high hydrostatic pressure at several hundred meters depth in the ice sheets, isolated air bubbles undergo a transformation into air-clathrate crystals (1) whose air-composition might differ from that of the original atmosphere (2, 3). In order to understand the processes which have influenced the compositional changes during the transformation of the system air-ice into clathrate, we have started a series of neutron powder diffraction and Raman scattering experiments which allow us to determine the filling isotherms of air-, N_2-, O_2-clathrates and fractionation coefficients of air-clathrates as a function of gas fugacity (4, 5). Our results have confirmed that the clathrate of air crystallises in the von Stackelberg type II cubic structure (S.G. Fd$\overline{3}$m) as previously shown in a single crystal X-ray diffraction study of a natural air-hydrate recovered from the Dye-3 drilling (Greenland)(6). The H-bonded, water, host lattice forms 16 small (SC) and 8 large (LC) cavities in which the guest molecules (N_2 and O_2) interact with the host through dipole-quadrupole electric interactions as well as dispersion forces. For a proper crystallographic modelling of the site disorder of the guest molecules, molecular dynamic (MD) simulations with subsequent quenches proved to be very helpful.

Experimental

Structural *in-situ* data on N_2-clathrates, obtained on the D2B diffractometer at ILL (Grenoble), were analysed by the Rietveld method using GSAS (7) to determine accurate values of the cage filling as a function of gas fugacity. Several conceivable crystallographic models were refined for a high quality data set using convergence behaviour and crystallographic agreement factors as selection criteria. It appears that only four disorder models were found relevant (labelled M1, M2, M3 and M4 in Table 1 & 2). The four models (M1, M2, M3, M4) differ only in the modelling of the small cage while the large cage is best described by only one model. Other models for the large cage led to unrealistic thermal parameters for the N_2 molecule. Incorrect thermal parameters may in turn by parameter correlations lead to incorrect occupancies (conf. Table 2 for M2 and M3). Thus, the Rietveld method is able to provide precise and accurate information on the atomic site occupancy only if the model used is the correct one. In order to find the appropriate model, we performed MD-simulations with subsequent quenches.

TABLE 1. Comparison of different models using convergence criteria of the refined neutron diffraction data on N_2-clathrate (synthesised at 523bar and 273K). Orientation of the molecular axis of the N_2 molecule, positional parameters and maximal site occupancy are given for each models.

M1	M2	M3	M4	M1, M2, M3, M4
small cage	small cage	small cage	small cage	large cage
[100]	[110]	[1 1̄ 0]	[110]+[1 1̄ 0]	[110]
(0,0,0.032)	(0.0225,0.0225,0)	(0.0225,−0.0225,0)	M2 + M3	(0.42,0.42,0.375)

TABLE 2. Results of the refinement using the models given in Table 1.

Results	Rwp	Rp	Filling SC (%)	Filling LC (%)	Uiso SC (Å²)	Uiso LC (Å²)
M1	2.36	1.5	91.2(6)	99.6(1.2)	0.0777(43)	0.174(12)
M2	2.58	1.69	92.4(6)	102.6(1.2)	0.0960(44)	0.204(12)
M3	2.59	1.64	94.8(6)	103.2(1.2)	0.1204(45)	0.189(13)
M4[a]	2.36	1.5	91.2(6)	99.6(1.2)	0.0796(44)	0.174(12)

[a] Occupancy as well as Uiso for both sites within the SC were constrained to be identical.

Simulation procedure

The quench MD-simulations[1] were performed using the MSI Cerius^2 suite of programs. The initial model, with its intermolecular parameters, is the same as the one used in our Monte-Carlo (MC) work (8) for the 523bar and 273K run but with full occupancy of the cavities. Intramolecular parameters for water are again the same as those used in our MC work (8) whilst the parameters for the nitrogen molecules were determined using our Raman data (5)

---

[1] Quench MD-simulations are periods of dynamics followed by a quench period in which the structure is minimized in energy at T = 0K.

on the basis of the Morse potential function. The lowest energy frame was extracted from the quenched trajectories and atom positions of the nitrogen molecules were translated from symmetry P1 into cubic Fd$\bar{3}$m. This was achieved by a computer program; a flow diagram of which is shown in Fig.1. Fig 2 shows the superimposed N_2 positions of the various quenches.

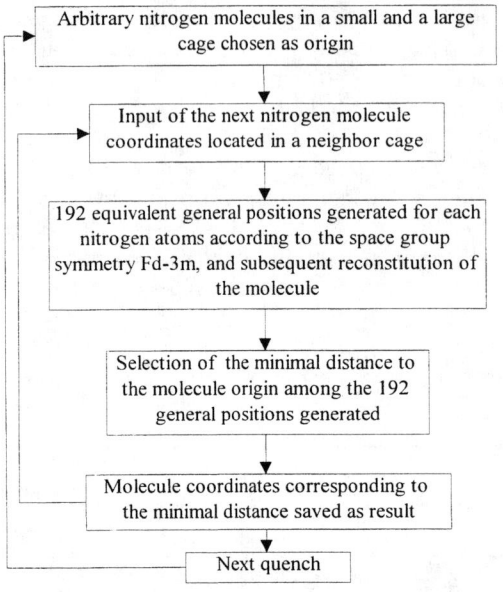

FIGURE 1 Flow chart of the transformation procedure from symmetry 1 to cubic Fd$\bar{3}$m to establish the most reliable crystallographic model for the small and large cage of type II clathrate.

It is noticable that the center of gravity for each N_2 molecule is loacated at the center of the SC, whereas the N_2 molecules in the LC are offset from the centre and sit closer to the cage wall. The N_2 molecular axis in the SC is found to point along the [100] direction. This distribution is in agreement with the M1 model proposed in table 1 (Fig.2.D.). Furthermore, the N_2 distribution in the LC also supports the M1 model for the LC with a molecular axis parallel to the direction [110] as it can be seen on Fig.2B, 2C. Most of the N_2 molecules (in light Fig 2B, 2D) can fit in one of the several positions available in the model. Orientations of molecules which differ slightly from that described by the M1 model, for both the small and the large cage, may be attributed to the expected local variations of the surrounding disordered water structure.

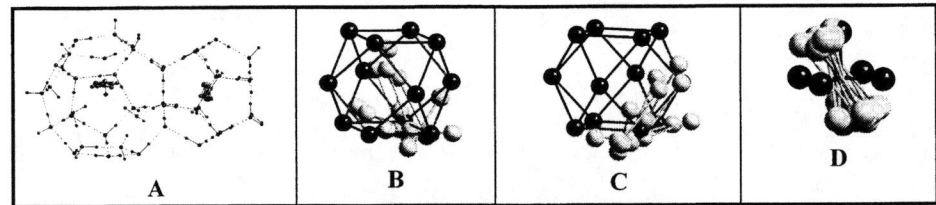

FIGURE 2. A. LC and SC with the transformed N_2 molecules from symmetry P1 to cubic Fd$\bar{3}$m. **B** and **C.** Model (black) in the LC compared with the transformed N_2 molecules (light molecules). **D.** Model M1(black) SC compared with the transformed N_2 (light molecules).

Comparison with neutron results

Difference Fourier maps between the observed and calculated scattering densities were used as an independent criterium for improving the quality of the refined crystallographic model. The difference maps for the four different models of the small cage (M1, M2, M3,

M4) are shown in Fig.3. The residual density in Fig. 3b, visible as two maxima and two minima, suggests that the real motion of the N_2 molecule within the small cage is anisotropic with largest thermal motion in the direction of the line adjoining the two maxima. Positive and negative residuals are also present in Fig.3c (M3) but distributed in a different way to that of Fig.3b. Residuals with lowest contrast are obtained for models M1, M2 and M4 (Fig.3a and 3d). A perfectly flat difference Fourier map cannot be expected as the guest molecules move in a highly anharmonic potential while the crystallographic model is based on a (harmonic) Gaussian density distribution. In spite of the relatively flat map obtained for M4, model M1 seems more appropriate regarding both the result of the quenches and the fact that U_{iso} of M4 is by no means smaller (despite a larger number of sites) than U_{iso} of M1. M2 appears to be less favoured than M1 considering the quality of fit (Table 2).

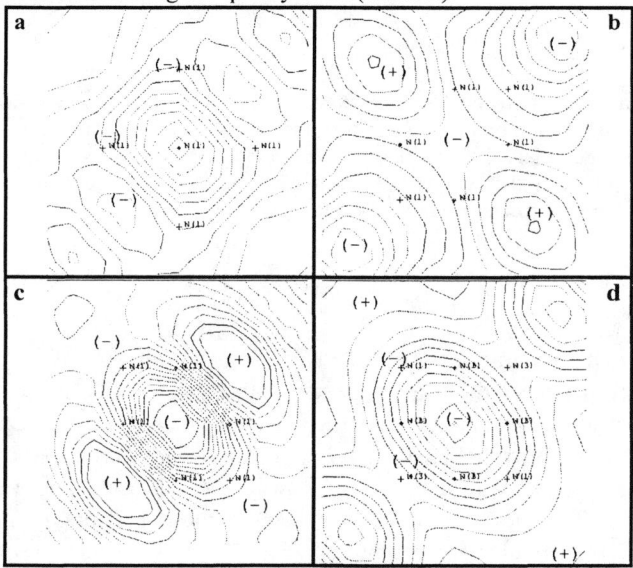

FIGURE. 3. Difference Fourier maps plotted in the (001) plane, in the vicinity of the small cage for N_2-clathrate synthesised at 523bar and 273K (**a** u.l. M1, **b** u.r. M2, **c** l.l. M3, **d** l.r. M4)

References

1. Shoji, H., Langway, C.C.Jr., *Nature* **298**, 548-550, 1982
2. Nakahara, J., Shigesato, Y., Higashi, A., Hondoh, T. and Langway, C.C.Jr., *Phil. Mag. B* **57**, 421-430, 1988.
3. Pauer, F., Kipfstuhl, S., Kuhs, W.F. and Shoji, H., *J.Glaciol.*, in press.
4. Kuhs, W.F., Chazallon, B., Radaelli, P.G. and Pauer, F., *J. Incl. Phenom.* **29**, 65-77, 1997.
5. Chazallon, B., Champagnon, B., Panczer, G., Pauer, F., Klapproth A. and Kuhs, W.F., *Eur. J. Mineral.* **10**, 1125-1134, 1998.
6. Hondoh, T., Anzai, H., Goto, A., Mae, S., Higashi, A. and Langway C.C.Jr., *J. Incl. Phenom.* **8**, 17-24, 1990.
7. Larson, A.C., von Dreele, R.B.: *GSAS-report LAUR* 86-748, 1-179, 1998.
8. Klapproth, A., Chazallon, B. and Kuhs, W.F., 1998, in Proceedings of „Neutrons & Numerical Methods", December 1998, Grenoble, France.

Neutron Scattering Studies of the Structure and Dynamics of Interlayer Water and Hydrated Cations in Montmorillonite Clays

D. Hugh Powell, Michael Gay-Duchosal and Cédric Pitteloud

Institut de Chimie Minérale et Analytique, Université de Lausanne-BCH, CH-1015 Lausanne, Switzerland.

Abstract. In order to understand the properties of smectite clays, we need to build up microscopic structural and dynamic models of the aqueous interlayer region between the negatively charged plate-like clay layers. We present a brief review of the role of neutron scattering in providing the quantitative information necessary for producing such models, with emphasis on our own recent results for Wyoming montmorillonite.

INTRODUCTION

Clays are the mineral component of soils with a mean grain size of the order of microns or less, and are probably the most important substrate for interactions between water, the mineral world and the biosphere. Ancient applications of clays include pottery and the clarification of beverages. Modern applications are found in paints, cosmetics, pharmaceuticals, foundry moulds, oil-well drilling-fluids, catalysis and environmental remediation (1, 2). The role of clays as containment barriers in radioactive waste repositories is of particular current interest (3).

Montmorillonite belongs to the class of clay minerals called smectites, which show extensive swelling on exposure to humidity, and is the essential component of the commercially important bentonite. Smectite particles are pseudo three dimensional crystals consisting of regular parallel stacks of negatively-charged aluminosilicate platelets held together by charge-balancing interlayer cations (typically Na^+ or Ca^+ in natural clays): Fig. 1. On exposure to humidity, the interlayer cations and the internal clay surfaces hydrate, causing the clay crystals to swell in a series of steps referred to as one-, two- and three- layer hydrates. The hydrated interlayer, and the exchangeable interlayer cations, are responsible not only for the swelling of smectites, but also for their exceptional ability to absorb a variety of charged, polar and non-polar species. A knowledge of the structural and dynamic properties of the interlayer region is thus essential for a full understanding of the behaviour of smectite clays in soils and in many of their applications.

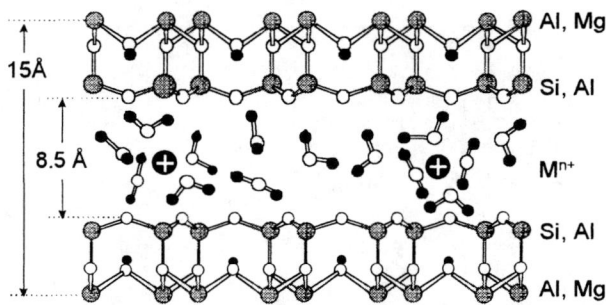

FIGURE 1. Cross-section perpendicular to the clay layers in Li-montmorillonite, with a hypothetical interlayer structure: O - oxygen, ● - hydrogen. The dimensions shown are for the two-layer hydrate.

RESULTS AND DISCUSSION

Earlier measurements (4) on partially oriented agglomerates of Na-montmorillonite showed that the interlayer D_2O makes a diffuse contribution to the overall neutron diffraction pattern. We performed measurements on powdered samples, in order to avoid the irreproducibility of the partial orientation of agglomerate samples. By using an approximate difference method, with the dry clay as a reference, we were able to extract the contribution due to the interlayer D_2O (5, 6). The example we show in Fig. 2 is taken from a recent measurement on the D4B diffractometer at the ILL. The top two traces are the corrected normalised intensities for vacuum dried Na-montmorillonite and the two-layer D_2O hydrate. If we make the approximation that the contribution of the crystalline diffraction from the clay platelets does not change on hydration, the difference function $\Delta(k)$ in Fig. 2(c) is a sum of partial structure factors corresponding to water-water and water-clay terms. The Bragg peaks due to the clay layers cancel, and we are left with a liquid-like diffraction pattern that resembles that of bulk D_2O. Fig. 3 shows the intermolecular radial distribution functions $G(r)$ obtained from $\Delta(k)$ for Wyoming montmorillonite hydrates with different interlayer cations. Comparison with the $G(r)$ for liquid D_2O shows that the interlayer water structure is similar to that in bulk water. There are differences in the region of 1.8 to 2.0Å that suggest differences in the hydrogen bonding structure between the different clay systems and compared to bulk water. We need, however, a more rigorous method to extract the interlayer diffraction. The method of isotopic substitution with first-order differences allows us to do this, and also to concentrate on the structure around the substituted species. We performed both H/D and $^6Li/^{nat}Li$ isotopic substitution measurements on the two-layer hydrate of Li-montmorillonite (7). The results showed clearly that water-clay interactions do not prevent the interlayer water from forming near the maximum of four hydrogen bonds per molecule and that the interlayer Li^+ has a well defined hydration shell, similar to that observed in aqueous solutions (8). This retention of bulk-like structure is impressive given the small thickness (*ca.* 8.5 Å) of the water layer in this clay.

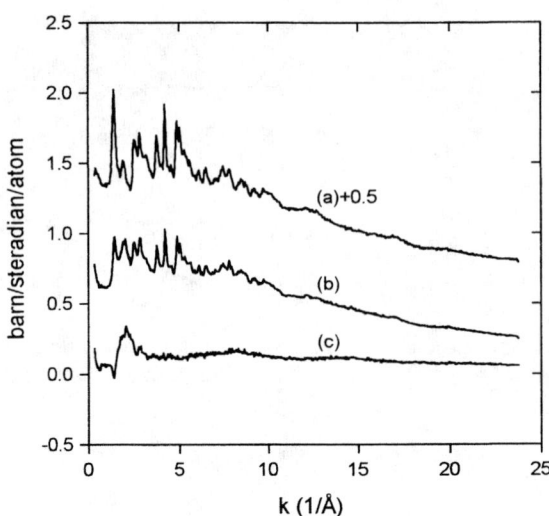

FIGURE 2. The corrected. normalised intensities (a) $I^{Dry}(k)$ for vacuum dried Na-montmorillonite and (b) $I^{D2O}(k)$ for D$_2$O hydrated (163 mg g^{-1} D$_2$O) Na-montmorillonite and (c) the difference function, $\Delta(k) = I^{D2O}(k) - 0.658\, I^{Dry}(k)$ (the factor 0.658 compensates for the lower clay fraction in the hydrated sample).

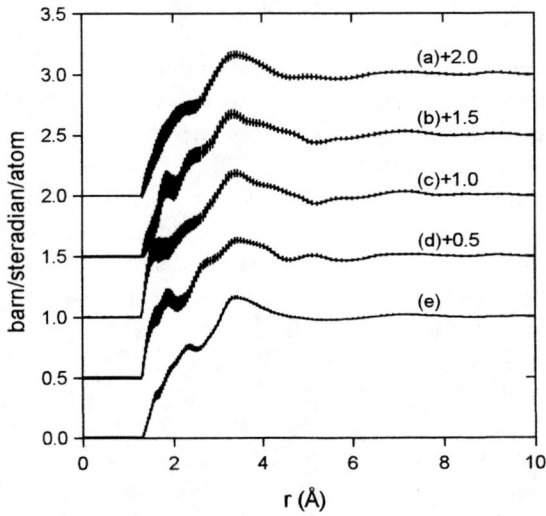

FIGURE 3. The intermolecular $G(r)$ for interlayer D$_2$O obtained after subtracting the dry clay pattern from that of the D$_2$O hydrated clay for (a) Li-montmorillonite (145 mg g^{-1} D$_2$O), (b) Na-montmorillonite (163 mg g^{-1} D$_2$O), (c) Mg-montmorillonite (200 mg g^{-1} D$_2$O) and (d) Ca-montmorillonite (201 mg g^{-1} D$_2$O) compared to (e) the $G(r)$ obtained for liquid D$_2$O under the same experimental conditions. The data have been normalised to unity at large r.

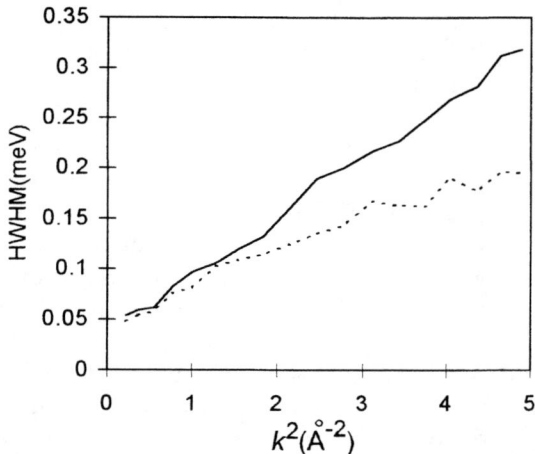

FIGURE 4: The quasielastic linewidth as a function of k^2 for the partially oriented three-layer hydrate of Na-montmorillonite obtained in transmission (line) and reflection (dashed line) geometry with an energy resolution of 100 µeV. For a scattering angle of $2\theta = 90°$ the two geometries correspond to k parallel and perpendicular to the clay layers respectively.

It is clear, therefore, that neutron diffraction with isotopic substitution can be used to obtain partial structure factors and radial distribution functions describing the liquid-like structure of interlayer water and hydrated cations in clays. Certain structural parameters (mean interatomic distances, coordination numbers) can be calculated directly from the data. However, in order to build up a detailed microscopic picture of the interlayer, and exploit the data to the full, it will be necessary to use some kind of modelling. The simplest application of the data in modelling is as a check of the physical reasonableness of Monte Carlo or Molecular Dynamics simulations. We will also explore, however, the possibility of using the data itself as a constraint for microscopic models using a method such as Reverse Monte Carlo (9). Since the crystalline structure of the clay layers themselves is not our primary interest, a reasonable starting point would be to treat the layer atoms as fixed and to constrain the positions of the interlayer cations and water using the relevant partial structure factors and combinations thereof. Since the cations may be associated with specific sites on the clay surfaces it may prove necessary, however, to allow in-plane movement of the clay layers. Since the interlayer spacing is not very well-defined by the broad (001) reflection in the total diffraction patterns, it may also be necessary to allow this parameter to vary, implying a variable box-size for the simulations.

Cebula et al (10) performed incoherent quasielastic neutron scattering (IQNS) measurements on partially aligned agglomerates of Li-montmorillonite hydrates, with an energy resolution of 41 µeV. Their transmission mode measurements were insufficient to detect anisotropy of the diffusive motion, but they were able to show that the translational diffusion coefficient is around 2, 3 and 5 times smaller than in bulk water in the three-, two- and one-layer hydrates respectively. We have recently performed IQNS measurements on partially oriented agglomerates of Na-montmorillonite on the NEAT (V3) time-of-flight spectrometer at the Hahn-Meitner Institut. As in the previous studies (10) the mosaic spread of the samples was rather large at around 40°. We measured in both transmission and reflection modes (k parallel and perpendicular to the clay platelets for a scattering angle of 90°) and at two energy resolutions (34 and 100 µeV). Fig. 4 shows that the quasielastic broadening for k parallel to the clay layers is greater than for k perpendicular to the clay layers, showing the anisotropy of the proton motion. A full model of the anisotropic motion as a function of scattering angle, taking into account the mosaic of the samples, will allow us to determine the detailed geometries and timescales of both the translational and rotational motions.

These results show that neutron scattering can furnish detailed information on the structure and dynamics of water and hydrated cations in the interlayer region of smectite clays. Our priority is to build microscopic models of the interlayer region for a series of typical smectite clays as a function of layer charge, water content and interlayer cation. We will explore the possibilities of parallel development of computer simulations with experiments, with the experimental results used as a quality control for the simulations, and of methods such as Reverse Monte Carlo, that allow detailed microscopic models to be built up directly from the experimental data.

ACKNOWLEDGEMENTS

We thank all those who have been involved in our neutron scattering work on clays, including C.J. Benmore, H.E. Fischer, T. Hauß, S.J. Kennedy, J.-C. Lavanchy, R.E. Lechner, P. Palleau, H.-R. Pfeiffer, B. Rufflé, N.T. Skipper, P.G. Slade, A.K. Soper and K. Tongkhao. We thank the Swiss National Science Foundation and the Federal Office of Education and Science (OFES) for their continuing financial support.

REFERENCES

1. Odom, I.E., *Phil. Trans. R. Soc. Lond. A* **311**, 171-189 (1984).
2. Bruce, D.W. and O'Hare, D., *Inorganic Materials*, Chichester: Wiley, 1992, ch. 6.
3. Austin, G.S., *New Mexico Geology*, 79-82 (1986).
4. Hawkins, R.K. and Egelstaff, P.A., *Clays Clay Miner.* **28**, 19-28 (1980).
5. Powell, D.H., Tongkhao, K., Kennedy, S.J. and Slade, P.G., *Clays Clay Miner.* **45**, 290-294 (1997).
6. Powell, D.H., Tongkhao, K., Kennedy, S.J. and Slade, P.G., *Physica B* **241-243**, 387-389 (1998).
7. Powell, D.H., Fischer, H.E. and Skipper, N.T., *J. Phys. Chem. B*, in press.
8. Howell, I. and Neilson, G.W., *J.Phys. Condensed Matter* **8**, 4455-4463 (1996).
9. McGreevy, R.L. and Pusztai, L., *Molec. Simul.* **1**, 359-367 (1988).
10. Cebula, D.J., Thomas, R.K. and White, J.W., *Clays Clay Miner.* **29**, 241-248 (1981).

Effect of charge transfer on the local order in liquid group IV isoelectronic compounds: neutron diffraction data versus numerical tight-binding simulations

G. Prigent*, J.-P. Gaspard[†], R. Bellissent* and C. Bichara[††]

*Laboratoire Léon Brillouin, CEA-CNRS, BP2, 91191 Gif-sur-Yvette cedex, France
[†] Condensed Matter Physique, University of Liège, B5, 4000 Sart-Tilman, Liège, Belgium
[††] Centre de Thermodynamique et de Microcalorimétrie, 26, rue du 141e R.I.A., 13003, Marseille, France

Abstract. In a simple tight-binding approach, we consider the role of charge transfer and entropy in the semiconductor-to-metal transition which may occur upon melting group IV elements and their isoelectronic III-V and II-VI compounds. In the liquid state, entropy is shown to destabilise the diamond structure in favor of a metallic simple cubic-like local order, while charge transfer tends to keep the semiconducting tetrahedral local order of the solid state. These results are consistent with neutron diffraction data.

INTRODUCTION

Structural trends within sp-bonded elements, as a function of the number of s and p electrons N_{sp}, have been extensively studied using Harrison [1] nearest-neighbour tight-binding models [2–5]. This approach, which can be easily extended to liquid and amorphous systems, has provided a simple and efficient insight into the study of the chemical bonding and the associated physical properties (bond length, compressibility...). For $N_{sp} < 4$, close-packed metallic structures were shown to be stable, but it was not possible to distinguish between FCC, HCP...etc. For $N_{sp} \geq 4$, the phenomenological Grimm-Sommerfield rule ($Z = 8 - N_{sp}$), used successfully for a large variety of covalent systems, was demonstrated analytically by Lannoo et al. [3] using a simplified density of states. Pettifor et al. [4] extended this study to all the elements taking into account the topology of local atomic environment through the behaviour of the first few moments of the densities of states, as suggested by Cyrot-Lackmann [6]. The agreement with experimental data was much more accurate. Recently, Gaspard et al. [7] showed that the structures of group V, VI and

VII elements and their isoelectronic compounds are a consequence of the Peierls instability [8]. Periodicity is not needed to obtain this instability, and consequently it may occur as well in crystalline, amorphous and liquid matter (at not too high temperature) which is in agreement with experiment [9–11].

In this paper we focus on the $N_{sp} = 4$ case, for which diamond structure leads to the formation of a sp^3 bonding mechanism. We discuss how the local order of liquid group IV elements or III-V and II-VI liquid is the result of a competition between charge transfer and entropic effects.

It is well known that group IV elements (Si, Ge...) and III-V compounds exhibit a common behaviour upon melting: their coordination number increases from four (diamond structure) to about six (simple cubic-like local order) [12,13], and they undergo a semiconductor-to-metal transition. Recent neutron diffraction data have shown that different behaviours appear upon melting II-VI compounds: a compound with a heavy cation (HgTe) undergoes the same structural change as group IV ements, whereas compounds with lighter cations (CdTe, ZnTe) keep the local environment of their crystalline phase and remain semiconductor in the liquid state [13–16]. We show here that this difference in behaviour is due to the large charge transfer in the latter.

The paper is organised as follow. In section 2 we present our tight binding model. In section 3 we evaluate the charge transfer effect on the semiconductor-to-metal transition in the solid state by comparing total energy of a simple cubic and a diamond structure for different values of the charge transfer. Finally, in section 4, we present some Monte Carlo simulation results that give some insight into the competition between charge and entropy in the iquid.

THE TIGHT-BINDING MODEL

The model used here is very similar to the model used by Bichara et al. for liquid As, Sb [17], Te [18] and Se [19]. The total energy is the sum of an attractive electronic energy, calculated in a tight-binding approximation, an empirical Born-Mayer repulsive energy and an electrostatic energy. Charge transfer is a difficult issue to address as it involves an empirical point of view. Many authors have proposed different possibilities for treating the Coulomb term in ionic compounds [1,20,21] (and more recently [22–24]). In a simple first approach, we used the Madelung point-like charge model [25].

The total energy E_t is then writen:

$$E_t = \frac{1}{2}\sum_{i \neq j} V_0 \exp(\frac{-pr}{r_0}) + \int_{-\infty}^{E_f} En(E)dE + \frac{-Q^2\alpha_s}{\frac{r}{r_0}} \quad (1)$$

where n(E) is the density of states (evaluated up to the fourth moment), E_f is the Fermi level, V_0 is fitted to have an equilibrium length of the simple cubic structure equal to r_0 (r_0, the length scale of our model, has been fixed arbitrary at 2.8Å which is the bond length of CdTe compound), Q is the charge transfer and α_s is the Madelung constant.

The tight-binding resonance integrals are assumed to vary with the distance following

$$\beta_\lambda = -\beta_\lambda^0 \exp \frac{-qr}{r_0} \qquad (2)$$

where the symbol λ denotes $ss\sigma$, $sp\sigma$ and $pp\sigma$ interactions ($pp\pi$ has been neglected). In order to reduce the number of parameters, we have taken from reference [1] the following canonical relation between the Slater Koster parameters β_λ^0:

$$\beta_{ss\sigma}^0 = -\beta_{sp\sigma}^0 = -\frac{1}{2}\beta_{pp\sigma}^0 \qquad (3)$$

Therefore, this model has four parameters: $\beta_{pp\sigma}^0$, ϵ_{sp}, p/q and Q. We have fixed an arbitrary energy scale with: $\beta_{pp\sigma}^0 = 4eV$.

CHARGE TRANSFER EFFECT

The semiconductor-to-metal transition occuring upon melting the group IV elements and III-V compounds is due to a structural transition between the semiconducting diamond structure towards a metallic six-coordinated disordered structure that is modelled by a simple cubic-like local order. Therefore, in order to get a better understanding of the role of charge transfer in the mechanism of the transition, we first compared at $T = 0K$ the total energy of simple cubic (SC) and diamond (DC) structures as a function of three parameters : $\epsilon_{sp} = \epsilon_p - \epsilon_s$, the ratio p/q (which characterises the hardness of the repulsion) and Q. The compared structures are relaxed: this is a NPT study.

No charge transfer ($Q = 0$)

We present in Fig. 1 our calculated structural map in the (ϵ_{sp}, p/q) space at $T = 0K$. We have also plotted the estimated positions of group IV elements (the values of ϵ_{sp} have been taken from [1]). It is shown here that, as expected, when the repulsive term gets harder (i.e. p/q increases), the SC structure is favoured. On the other hand, for a small value of ϵ_{sp}, it is possible to stabilise sp^3 hybrids. As the $pp\sigma$ bond is more "deformable" than sp^3 bond (the function $E_t(r)$ has a smaller curvature for the diamond structure at equilibrium bond length than for the SC structure), it is intuitive than for $T \neq 0K$, the entropic effect will favour the SC structure, so that the curve will be shifted toward the small values of p/q, which would explain the structural transition observed at melting. In the next section, we present a Monte Carlo simulation which confirms this hypothesis.

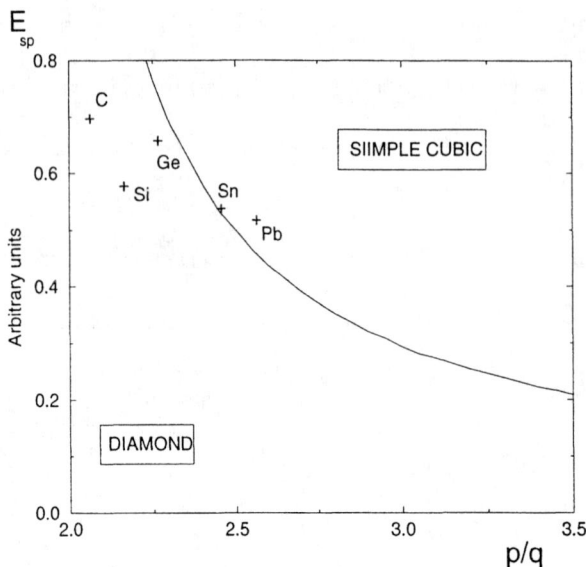

FIGURE 1. Structural map in the $(p/q, \epsilon_{sp})$ space without charge transfer. Diamond and simple cubic structures are compared at equilibrium bond length for $T = 0K$. We have plotted the estimated position of the group IV elements.

With charge transfer ($Q \neq 0$)

We have plotted on Fig 2 the calculated map for different values of Q at $T = 0K$. A critical value p/q^* arises: for $p/q < p/q^*$, charge transfer favours DC structure whereas for $p/q > p/q^*$ the opposite behaviour is observed. This crossover depends on the value of the equilibrium bond length of both structures. As a matter of fact, we can roughly say that for $p/q < p/q^*$, we have $\frac{r_{DC}}{r_{SC}} < \frac{a_{DC}}{a_{SC}}$. Moreover, it is remarkable that the critical value p/q^* observed here is the same as that which distinguishes between Peierls distorted and undistorted structures [7].

ENTROPIC EFFECTS

In the previous section, we suggested that the sp^3 hybrid bonding found in group IV elements can be destabilized by entropic effects in favour of a non-hybridised $pp\sigma$ interaction (the solid line of Fig. 1 is shifted toward small value of p/q for $T \neq 0K$). In order to check this hypothesis, we performed a NPT Monte Carlo simulation ($T \neq 0K$) with the parameters of a point belonging to the diamond structure favourable phase (star on Fig. 2). We have plotted on Fig. 3 the pair correlation function of the corresponding liquid system. The coordination number increased from 4 in the solid state to about 6.1 in the liquid, and the ratio of the

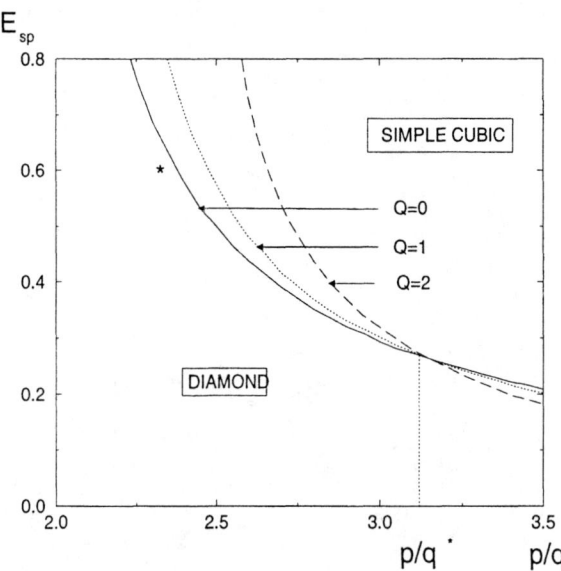

FIGURE 2. Structural map in the $(p/q, \epsilon_{sp})$ space with different values of the charge transfer. Diamond and simple cubic structures are compared at equilibrium bond length for $T = 0K$. The star indicates the chosen point for the Monte Carlo simulations.

second-to the first-neighbour distances gave a bond angle of 89, indicating that the local order is close to a simple cubic structure. These results are in good agreement with diffraction experiments. Recent neutron diffraction data have shown that some II-VI compounds keep the tetrahedral local environment of their crystalline phase in the liquid state [14,15]. The results of the NPT Monte Carlo simulations with $Q \neq 0$ will be publish elsewhere.

CONCLUSION

We have used a tight-binding approach to evaluate the importance of charge transfer in the structural transition between a diamond structure and a simple cubic structure, modeling the two structures in competition during the melting of group IV and their isoelectronic III-V and II-VI compounds. When the repulsive potential is not too hard ($p/q < p/q^*$), the charge transfer is shown to favour the diamond structure in the solid state. In the liquid state, charge transfer may favor tetrahedral local order as long as entropic effects are not too strong. These results are in good agreement with neutron diffraction experiments performed on liquid CdTe and ZnTe.

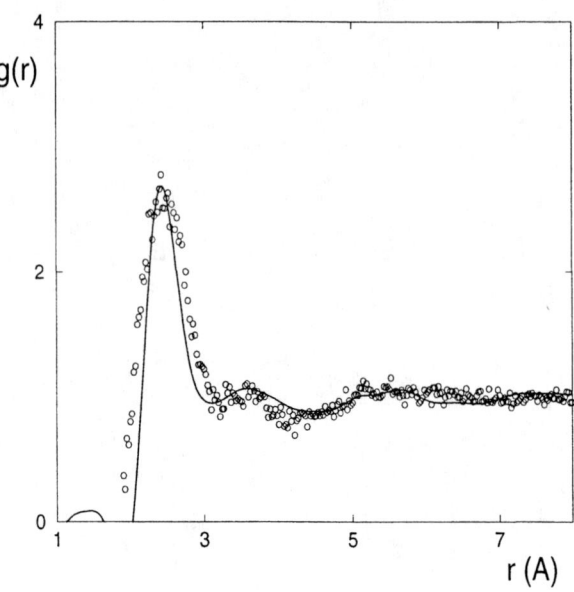

FIGURE 3. Calculated pair correlation function in the liquid state, just above the melting point, for $\epsilon_{sp} = 0.6$ a.u. and $p/q = 2.3$ (circles). In addition we have plotted the experimental data for l-Si. The diamond structure is destabilised in favor of a simple cubic like local order.

REFERENCES

1. W.A. Harrison, *Electronic structure and the properties of solids*, San Francisco, Freeman,1980
2. F. Ducastelle, *J. de Physique* **31** 1055 (1970)
3. G. Allan and M. Lannoo, *J. de Physique* **44** 1355 (1983)
4. J.C. Cressoni and D.G. Pettifor, *J. Phys.: Condens. Matter* **3** 495 (1991)
5. J. Freidel, *J. de Physique* **39** 651 (1978)
6. F. Cyrot-Lackmann, *Adv. in Phys.*, **16** 393 (1967)
7. J;-P. Gaspard, A. Pellegatti, F. Marinelli and C. Bichara, *Phil. Mag. B*, **77** (3) 727 (1998)
8. R. E. Peierls, *Quantum theory of Solids* Oxford Univ. Press, 1955
9. R. Bellissent and G. Tourand, *J. Phys. Paris* **37** 1423 (1976)
10. M.F. Daniel and A.J. Leadbetter, *Phil. Mag. B* **44** 509 (1981)
11. J.-P. Gaspard, R. Bellissent, A. Menelle, C. Bergman and R. Ceolin, *Nuovo Cimento* **12** 649 (1990)
12. Y. Waseda, *The structure of Non-Crystalline Materials. Liquid and amorphous Solids* New York, Mc Graw-Hill, 1981
13. C. Bergman, C. Bichara, P. Chieux and J.P. Gaspard, *J. Phys. (Paris) C* **46** (8) 97 (1985)

14. J.P. Gaspard, J.Y. Raty, R. Céolin, R. Bellissent, *J. of Non-Cryst. Solids*, **75** 205 (1996)
15. G. Prigent, R. Bellissent, R. Céolin, H.E. Fischer, J.P. Gaspard, *J. Non-Cryst. Solids*, in press
16. J.-Y. Raty, j.-P. Gaspard, R. Céolin and R. Bellissent, *Physica B* **234-236** 364 (1997)
17. C. Bichara, A. Pellegatti, J.P. Gaspard, *Phys. Rev. B* **47** 5002 (1993)
18. C. Bichara, J.Y. Raty, J.P. Gaspard, *Phys. Rev. B* **53** 206 (1996)
19. C. Bichara, J.Y. Raty, J.P. Gaspard, submitted in *Phys. Rev. B*
20. J.C. Phillips, *Bonds and Bands in Semiconductors*, New York, Academic Press, 1973
21. J.A. Majesvky and P. Vogl, *Phys. Rev. B* **35** 9666 (1987)
22. R.D. King-Smith and D. Vanderbilt, *Phys. Rev. B*, **47** 1651 (1993)
23. M. Di Ventura and P. Fernandez, *Phys. Rev. B* **56** R12698 (1997)
24. J. Bennetto and D. Vanderbilt, *Phys. Rev. B* **53** 15417 (1996)
25. E. Madelung, *Gott. Nach.* **100** 304 (1909)

Geometric frustration in gadolinium gallium garnet: a Monte Carlo study

Oleg A. Petrenko and Don McK. Paul

University of Warwick, Department of Physics, Coventry CV4 7AL, UK

Abstract. We have studied the magnetic properties of the frustrated triangular antiferromagnet $Gd_3Ga_5O_{12}$ (GGG) by means of classical Monte Carlo simulations. Low-temperature specific heat, magnetization, susceptibility, autocorrelation function and neutron scattering function have been calculated for several models including different types of magnetic interactions and with the presence of an external magnetic field. In order to reproduce the experimentally observed properties of GGG, the simulation model must include nearest neighbor exchange interactions and also dipolar forces. In zero field there is a tendency to form incommensurate short-range magnetic order around positions in reciprocal space where antiferromagnetic Bragg peaks appear in an applied magnetic field.

I INTRODUCTION

The introduction of frustration to magnetic systems leads to extra degeneracy of the ground state in addition to the degeneracy resulting from the symmetry of magnetic Hamiltonian. The larger this additional degeneracy, the more likely frustration is to cause dramatic changes in the magnetic properties of the system, such as an absence of long range ordering even at the lowest temperature. Although geometrical frustration has been one of the key issues in magnetism for the past twenty years, only recently has it been established that the additional degree of freedom caused by frustration could be extensive, as in case of pyrochlores, where it is proportional to the number of spins involved [1], or in the case of *Kagome* lattices.

In gadolinium gallium garnet, $Gd_3Ga_5O_{12}$, (space group $Ia\bar{3}d$) the magnetic Gd ions are located on two interpenetrating, corner-sharing triangular sublattices, where the triangles of spins do not lie in the same plane – the angle between two nearest triangles is equal to the angle between the diagonals of a cube, 73.2° (see Fig. 1). In this compound the triangular arrangement of the nearest spins is combined with a complete exchange isotropy and with a relatively strong dipole-dipole energy. Although the magnetic properties of various garnets have been thoroughly studied in the past, the analogy between any of them and GGG is not straightfor-

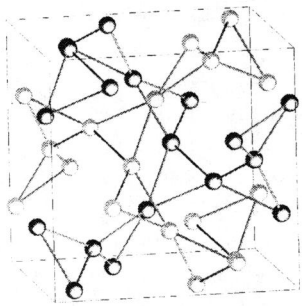

FIGURE 1. Positions of the magnetic Gd ions in a garnet structure. There are 24 magnetic ions per unit cell, they are divided into two interpenetrating sublattices.

ward: all other magnetic garnets order below 1 K, while no long range magnetic order has been detected in GGG down to 25 mK [2].

In this paper we present the results of classical Monte Carlo (MC) simulations of the magnetic properties of GGG. MC simulations provide an adequate alternative to theory, they also bring an additional depth to the understanding of experimental results obtained for geometrically frustrated magnetic systems. The original idea to simulate the magnetic properties of GGG using the MC method belongs to Kinney and Wolf [3], who calculated the temperature dependence of the specific heat and by comparing the results with experimental data obtained the amplitudes of the nn and nnn interactions J_1, J_2 and J_3. We use the same value of the nn exchange constant, $J_1 = 0.107$ K, however, as it will be shown later, the values of J_2 and J_3 quoted in [3] are not essential: as long as they are small in comparison with J_1, they do not change significantly the calculated magnetic properties of GGG and therefore can not be reliably determined from MC simulations. We also discuss some important questions such as the minimum simulation time required to reach thermal equilibrium, the influence of the dipole-dipole cut-off range and role of the size of the simulated lattice.

II SIMULATION MODELS

MC simulations have been performed for lattice sizes $L \times L \times L$, with $L = 3$ to 9 unit cells, containing 648 to 17496 spins. Significantly larger lattice sizes ensured that the magnetic correlation length in the disordered phase does not exceed the system size. They also have improved the resolution of the calculated scattering function, $S(Q)$, in a magnetic field allowing one to resolve clearly individual magnetic Bragg peaks. A standard Metropolis algorithm with periodic or open boundary condition has been employed; up to a million Monte Carlo steps per spin (MCS) were performed at the lowest temperatures. An attenuation factor δS has been introduced in such a way that roughly 50% of the attempted spin moves were accepted [4], which has resulted in a dramatic increase of the spin relaxation rate. For the simulations in a magnetic field this procedure have been abandoned in order to permit the system to make abrupt rearrangements.

It is very important and at the same time very difficult to simulate reliably such a long range interaction as the dipole-dipole one. Previous simulations have restricted the dipole-dipole interaction to third neighbor [3,5], while in our model it has been extended to include the fourth neighbor. We have also made several test runs to compare the simulation results for this model with the previously used cut-off range and have found no significant difference, which suggests that this model describes reasonably well the contribution of the dipolar force.

III RESULTS AND DISCUSSION

A. Zero external field properties.

We start with addressing the issue of a possible transition to a spin-glass state at $T = 140$ mK [6]. A model, which includes only nn exchange, J_1, does not show any indication of a phase transition down to at least $T = 2$ mK (which is less than 0.2% of JS^2). The system remains in a spin-liquid phase: averaging over sufficiently long time gives zero magnetic moment on each site, the magnetic correlation length does not exceed the unit cell size, there is no maximum or a cusp in heat capacity as a function of temperature, the scattering function does not show any sharp peaks.

The introduction of dipolar interactions slows down the spin-relaxation process. Fig. 2 displays time dependence of the autocorrelation function $A(t) = \frac{1}{N}\sum <\mathbf{S}_i(0)\mathbf{S}_i(t)>$ for the two models: with (bottom) and without (top) dipolar forces. Note, that the difference between the autocorrelation function for these two models is evident at all temperatures below 0.5 K, which approximately coincides with the nn dipole-dipole energy.

When dealing with very slow spin systems and a potential spin-glass phase transition it is most important to ensure that the simulation time is longer than the necessary equilibration time. In practice the first t_0 MCS are used only for equilibration and then calculations and averaging are carried out during the next t_0 steps. In a GGG model which includes both exchange and dipolar interactions t_0 becomes enormously long at temperature below $T = 100$ mK. Therefore the results of calculations in a zero field for a model which includes dipole-dipole interactions could not be considered as reliable below this temperature. The problem of long equilibration times is removed by breaking symmetry through the application of an external magnetic field.

The results of the simulations with the model which takes into account only the nn exchange interaction, J_1, fit nicely the experimental neutron scattering function $S_p(Q)$ at all temperatures above 140 mK [7]. The neutron scattering function $S_p(Q)$ for a powder sample is calculated from:

$$S_p(Q) = f(Q)^2 \sum_{i,j} <\mathbf{S}_i\mathbf{S}_j> \frac{Sin(Qr)}{Qr}, \qquad (1)$$

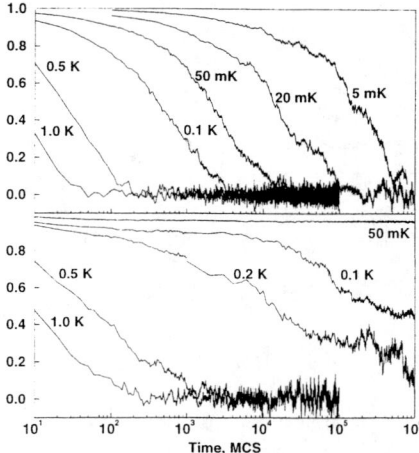

FIGURE 2. Time dependence (in MCS per spin) of the autocorrelation function $\frac{1}{N}<S(0)S(t)>$ for various temperature in a model which includes: only the nn exchange, J_1 (top), the nn exchange and the dipole-dipole interactions (bottom). System size is $4\times4\times4$ unit cells.

where $f(Q)$ is the magnetic form factor. The intensity of the broad diffuse scattering peaks increases as the temperature decreases in total agreement with the experiment. The introduction of the dipole-dipole interaction at these temperatures does not change $S_p(Q)$ significantly.

Somewhat unexpected results have been obtained for a single crystal neutron scattering function, calculated by:

$$S_{xt}(Q) = (f(Q)\sum_{n}^{N} \mathbf{q}_n e^{i\mathbf{Q}\mathbf{r_n}})^2, \qquad (2)$$

where $\mathbf{q}_n$ is the magnetic interaction vector. Well above $T = 140$ mK, where there is no problem with very long equilibration times, $S_{xt}(Q)$ demonstrates a tendency to form incommensurate peaks around integer positions in the reciprocal space (see Fig. 4). The intensity of these incommensurate peaks is much lower than the expected intensity of the true long-range order Bragg peaks, and their width is determined by the system size.

An interesting question to investigate is how the ratio of exchange to dipolar interactions influences the magnetic properties of GGG. In GGG the nn exchange, J_1, is about twice the strength of the nn dipolar interactions, DD, and there is no magnetic order, while in the Mn-based garnets [8] the ratio J_1/DD is slightly higher and they do order. For instance, in $Mn_3Al_2Ge_3O_{12}$, which undergoes an antiferromagnetic phase transition to a $120°$-structure at $T_N = 6.65$ K, $J_1 = 0.57$ K [9] is more than ten times stronger that DD. Could this fact along lead to an appearance of long-range magnetic order? Our simulation results do not support this idea: in a model, where J_1 has been increased up to a hundred times keeping the DD value fixed, the ground state remained disordered. However, the introduction

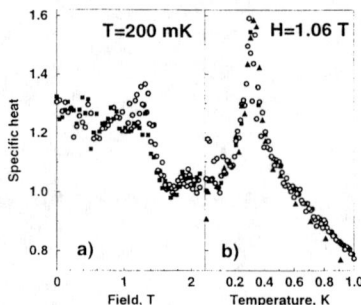

FIGURE 3. a) Specific heat field dependence for a $3\times3\times3$ model which includes J_1 and the dipolar interactions (open symbols) and for a model which includes J_2 and J_3 as well (solid symbols). b) Temperature dependence of the specific heat for a model which includes J_1 and dipole-dipole interaction. 10^5 (open circles) and 10^6 (solid squares) MCS have been performed.

of the nnn exchange interaction with a value cited in [9], $J_2 = 0.12$ K, does make a difference: the system undergoes a phase transition to a LRO state, which reveals itself clearly both as a cusp in the heat capacity temperature dependence and as peaks in the scattering function, $S(Q)$.

B. Magnetic properties in an applied field.

As it has been mentioned above, in an applied magnetic field the problem of long equilibration times is much less severe, which gives us an excellent opportunity to investigate the magnetic phase diagram of GGG in detail. The phase transition to a LRO state in magnetic field was detected by calculating the specific heat temperature dependence in constant field or by calculating its field dependence at constant temperature. In order to avoid problems with possibly multiple metastable states the calculations were always started at higher temperatures and fields and gradually shifted to the appropriate conditions. Fig. 3 gives the examples of such calculations. The position of the specific heat maximum is not sensitive to the introduction of relatively weak nnn exchange interactions (such as these quoted in [3], $J_2 = -0.003$ K and $J_3 = 0.010$ K), neither does it show any visible size-dependence. In the field dependence of the specific heat, only one anomaly corresponding to the upper transition field is well-pronounced, while there is no obvious anomaly corresponding to the lower transition field, which agrees with previous MC simulations [5]. Overall the position of the specific heat peak agrees well with the experimental phase diagram of GGG.

In an applied magnetic field, where LRO is developed, the formula (1) is not valid any more: although it indicates clearly the appearance of magnetic Bragg peaks, their intensity is not calculated correctly. However the overall field dependence of the intensity mimics extremely well the experimental data [10]. There are two different groups of magnetic peaks, ferromagnetic and antiferromagnetic. The intensity of the former group is growing in lower field and saturating in higher field. The intensity of the latter group also grows in lower field reaching its maximum at around $H = 1$ T and then decreases in higher fields and disappears above $H = 2$ T.

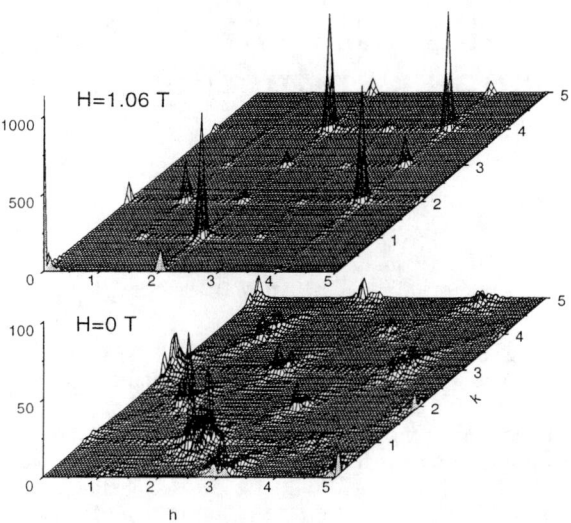

FIGURE 4. Simulated single crystal neutron scattering intensity of GGG in the ($hk0$) plane at $T = 0.2$ K in a zero field (bottom) field and in a field of $H = 1.06$ T applied along (001) direction. The data have been obtained by averaging 100 runs 2500 MCS each (bottom) and 10 runs 10000 MCS each (top) for a lattice size $9 \times 9 \times 9$ unit cells.

Exactly the same behavior has been seen by simulating single-crystal scattering intensity according to formula (2). Fig. 4 shows the results of such calculations for $H = 1.06$ T and $T = 200$ mK. It should be emphasized that no antiferromagnetic peaks have been observed in a model which excludes the dipolar forces.

In conclusion, we have extended previous MC calculations on GGG in order to provide a better comparison with our experimental data. Experimental properties of GGG including the low temperature magnetic phase diagram are consistent with our findings. The presence of incommensurate short range order in zero magnetic field predicted by the simulations needs experimental verification.

REFERENCES

1. Moessner R., and Chalker J. T., *Phys. Rev. B* **58**, 12049 (1998).
2. Ramirez A.P., and Kleiman R. N., *J. Appl. Phys.* **69**, 5252 (1991).
3. Kinney W. I., and Wolf W. P., *J. Appl. Phys.* **50**, 2115 (1979).
4. Reimers J. N., *Phys. Rev. B* **45**, 7287 (1992).
5. Schiffer P et al., *Phys. Rev. Lett.* **73**, 2500 (1994).
6. Schiffer P et al., *Phys. Rev. Lett.* **74**, 2379 (1995).
7. Petrenko O. A. et al., *Phys. Rev. Lett.* **80**, 4570 (1998).
8. Prandl W., *Phys. Stat. Sol. (b)* **55**, K159 (1973).
9. Valyanskaya T. V. et al., *Sov. Phys. JETP* **43**, 1189 (1976).
10. Petrenko O. A. et al., to appear in *Physica B* (1999).

Molecular Modelling of Organic Superconducting Salts

S.A. French and C.R.A. Catlow

The Royal Institution.
21 Albemarle Street, London, W1X 4BS

Abstract. We have undertaken a detailed computational study of the organic superconducting charge transfer BEDT-TTF salts. The salts have wide ranging physical properties which are derived primarily from the packing of the donor-radical cations. Electronic structure calculations have been performed at the molecular and periodic level using Density Functional Theory (DFT). These calculations have been used to investigate the magnetic ordering effect of the BEDT-TTF $FeBr_4$ salt. By fixing orbitals occupation we were able to compare relative energies of various spin states. The intermolecular spin ordering was found to be primarily mediated via the covalent sulphur-sulphur interaction on neighbouring BEDT-TTF cations. The structure prediction of the BEDT-TTF salts presents a serious challenge to full QM calculations. As an alternative method, molecular mechanics was employed to model the structures of a number of large molecular crystals. Results using the ESFF forcefield in the Molecular Mechanics (MM) code, Discover, are presented.

INTRODUCTION

BEDT-TTF [bis(ethylenedithio)tetrathiafulvalene] or "Et", first synthesised by Mizuno et al. [1] has been extensively studied due to it being the organic component of a family of superconducting molecular charge transfer salts. An important structural aspect of these systems, based on the donor molecule Et, is the spatial segregation of the organic cations and the inorganic anions into alternating layers and stacks. As a consequence, there are many lattice packings of Et, which lead to insulating, semiconducting, metallic and superconducting phases. These physical properties result primarily from the packing of the donor-radical cations in the crystal structure. The orientation and distance between the cations is critical to the electronic behaviour. The sulphur-sulphur interaction is the major influence on the intermolecular interactions due to the larger size of these heteroatoms. To gain an insight into the structure and electronic properties, calculations have been performed using both Molecular Mechanics (MM) and electronic structure techniques. MM is especially suitable for investigations of structure and stabilities of crystals. The electronic structure of the cation is also addressed using molecular Quantum

Mechanical (QM) techniques. QM periodic calculations have been performed to investigate one electron spectra and magnetic ordering for superconducting salts using as an example EtFeBr$_4$ [2].

METHODS

MM undertakes conformational analysis using a forcefield based method at a fraction of the computational cost of a QM calculation. As the systems being studied range from 68 to 500 plus atoms in a unit cell, the availability of a classical method is essential. Due to the diverse nature of atoms and particularly metals in the crystals, a forcefield is required that has a wide range of interatomic potentials. The Extensible and Systematic forcefield (ESFF) [3], which was designed for modelling organic, inorganic and organometallic systems was used together with Discover [3], which employs standard minimisation techniques. To reduce the number of required potentials, the forcefield employs semi-empirical rules to translate atomic based parameters to interatomic parameters typically associated with a covalent forcefield. As crystals are being modelled, periodic boundary conditions and an Ewald summation for electrostatic interactions are used.

Electronic structure calculations have been performed using Density Functional Theory (DFT). Molecular calculations were performed using DMol [4] and DMol^3 [5] while DSolid [6] and DMol^3 [5] were used for periodic calculations. All codes use numerical basis sets where the basis functions are given as values on an atomic centred spherical polar mesh. Calculations were performed using Double Numeric (DN) and Double Numeric + Polarisation (DNP) basis sets. Calculations have been performed using both local, Perdew and Wang 1981-1991 [7] and non-local Becke 1988 [8] + Perdew Wang 1991 [7] and Perdew Wang 1991 [7] functionals.

RESULTS

Molecular Mechanics

Before any prediction of new structures can be made, the ESFF forcefield, must be rigorously tested on known crystal structures. The two systems chosen were (i) Et neutral containing 104 atoms in a unit cell - one of the starting reactants in the synthesis of Et salts and (ii) the 1:1 salt EtFeBr$_4$ containing 68 atoms. Root Mean Square (RMS) deviations of atomic positions, comparing observed and calculated crystal structures, were used to compare different combinations of inter-atomic potentials. The best result for the neutral salt was an RMS of 0.298$Å$; comparison of experiment and calculation is illustrated in fig1.

EtFeBr$_4$ has a triclinic unit cell and contains two cations and two tetrahedral FeBr$_4$ anions, the iron centre being formally 3+. The main feature of the structure of EtFeBr$_4$ is the absence of stacks and planes of closely spaced Et molecules,

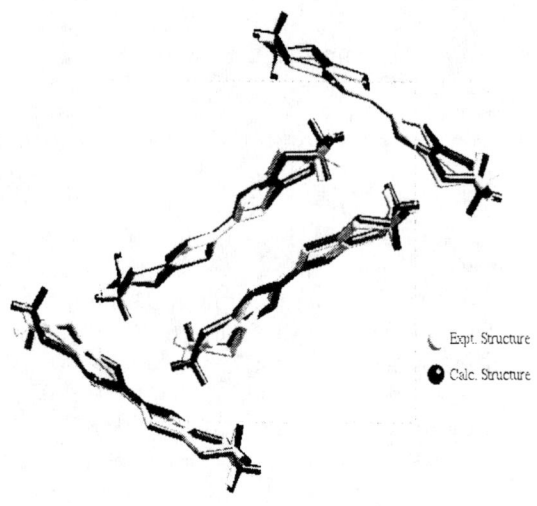

FIGURE 1. Comparison of observed and calculated crystal structures.

in marked contrast to most Et compounds. The Et molecules are fully oxidised, having a formal charge of +1 and are much more planar that those in the neutral salts (as discussed further below when we descibe the periodic electronic structure calculations). The ESFF forcefield still manages to model this subtle change in crystal field effectively. The RMS value is 0.409Å. The problem for modelling more complex salts such as BEDT-TTF$_2$FeCl$_4$ [2], is that the ratio of cations to anions leads to non-integer formal charges. As the MM procedures within Discover do not allow non-integer formal charges the electrostatic contributions will be incorrect. A modified version of ESFF [3] has been used which allows the manual input of partial charges. The RMS deviation for Et$_2$FeCl$_4$ is 0.396Å. The largest system studied was Et$_2$I$_3$ [9] with 110 atoms per unit cell and an RMS deviation of 0.228Å. Overall, the results show that ESFF type forcefields can model the crystal structures of the compounds with acceptable accuracy.

Molecular DFT Calculations

We now investigate using DFT techniques, the electronic structures of both isolated Et molecules (and ions) and of the Et salts. The electronic structure of a single Et molecule has been explored before and after ionisation to model the process of charge transfer which is controlled by the Ionisation Potential (IP) of the cations and the electron affinity of the the anions. Upon optimisation with no symmetry there is a large structural change between the neutral species and the more planar cation. The results agree well with other published work using Extended Huckel Theory (EHT) [10] and Hartree Fock (HF) [11]methods. The Highest Occupied Molecular Orbital (HOMO) of Et has been found to be of carbon double bond (C=C) character for bonding and carbon-sulpur single bond (C-S) for antibonding. Upon ionisation, the bonding character of the central C=C decreases leading to an increase in bondlength from 1.34Å to 1.39Å and 1.34Å to 1.38Å for the outer C=C. All carbon-sulphur bondlengths decrease. The IP of a single Et molecule has been calculated in two ways. The energy of the Et^0 molecule can be minimised and a single point calculation undertaken with a charge of +1 according to the Franck Condon principle, giving the vertical IP, ΔE_0. The other method compares fully optimised structures for both charge states 0 and +1, ΔE_1. Our results give ΔE_0=6.57eV, ΔE_1= 5.85eV, which compares to 6.5eV and 6.21eV respectively from experiment [12] and 6.87eV and 5.77eV for calculations performed at the HF/6-31G** level by Demiralp et al [11].

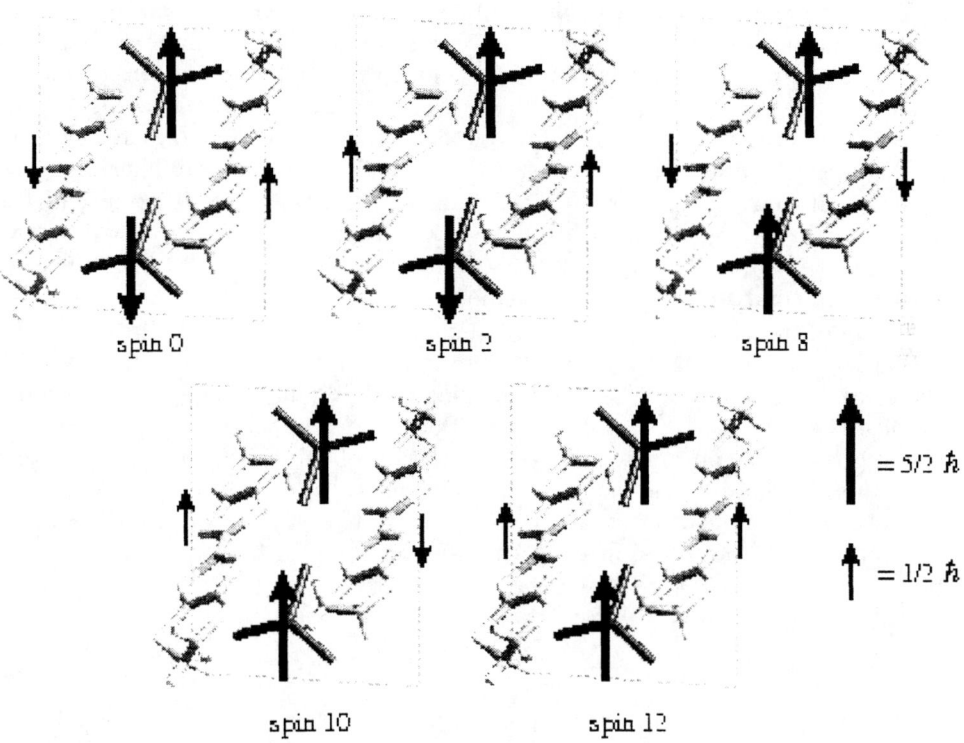

FIGURE 2. The 5 possible spin states of $EtFeBr_4$.

TABLE 1. Energies of different spin states. All results are for single point calculations using DN basis set and the Perdew 1991 functional for exchange and correlation. (Δ is the difference between excited and ground state.)

Spin State	Total Energy in Ha	ΔeV	Δkcal/mol
0	-30264.91437	—	—
2	-30264.90042	0.37960	8.75370
12	-30264.89545	0.51484	11.87239

Periodic DFT Calculations

Periodic calculations have been performed on EtFeBr$_4$ using DFT to include both steric constraints and crystal field effects in our model. These calculations are computationally very expensive, so only single point calculations have been performed so far. The mechanism of superconductivity is unknown but clearly charge transfer is vital. The formation of stacks leads to holes being present at an intermolecular distance of approximately 3.6Å which has led to the hypothesis that carriers hop between Et molecules via the sulphur-sulphur intermolecular interactions. EtFeBr$_4$ is an insulating phase due to the formation of dimers, with sulphur-sulphur distances of 3.6Å, with no pathway for the hopping of conduction electrons due to a barrier of anions. EtFeBr$_4$ can still possess short range electronic pairing interactions characteristic of many superconductors but not have any of the long range interactions. Of interest is the spin state of the molecule. From Mulliken analysis it is clear that charge transfer does occur and that each Et has a +1 oxidation state. The Fe centre is high spin and d^5 as seen in experimental results. There are five different possible spin states 0, 2, 8, 10, 12 (in units of $\frac{1}{2}\hbar$). The energies of the systems are compared to find the ground state.

The predominant force controlling magnetic ordering is the intermolecular exchange-correlation interactions rather than that in a simple ionic material where the controlling force would be the Madelung field. We have encountered technical problems for the 8 and 10 spin states. Both states appear to be unstable with respect to a full (spin unconstrained) relaxation of the wavefunction of EtFeBr$_4$, after a particular starting value of spin has been introduced. Hence we conclude that the stable spin states are 0, 2 and 12. Work is in progress to isolate 8 and 10 spin states with the aim of investigating the magnetic exchange energies.

CONCLUSIONS

Our results demonstrate the applicability of MM to further investigation of unknown crystal structures while the QM calculations provide a tool for further insight into the electronic properties both of the constituent molecules and their salts.

ACKNOWLEDGEMENTS

We are grateful to EPSRC for financial support. We would like to thank MSI for the provision of all the software used and partial support of the project. We are grateful to Prof P. Day for advice and helpful discussions.

REFERENCES

1. Mizuno M., Gariot A., Cava M., *J. Chem. Soc. Chem. Commun.* 18 (1978).
2. Mallah T., Hollis C., Bott S. Kurmoo M., Day P., *J. Chem. Soc. Dalton Trans.* 859 (1990).
3. Discover96.0/4.0.0, September 1996. San Diego: Molecular Simulations, (1996).
4. DMol96.0/4.0.0, September 1996. San Diego: Molecular Simulations, (1996).
5. Cerius^{2}3.5, September 1997. San Diego: Molecular Simulations, (1997).
6. DSolid 4.0.0, September 1996. San Diego: Molecular Simulations, (1996).
7. Perdew, J. P., Wang Y., *Phys. Rev.* **B54**, 13244 (1992).
8. Becke, A. D., *J. Chem. Phys.* **88**, 2547 (1988).
9. Bender K., Hennig I., Schweitzer D., Dietz K., Endres H., Keller H. J., *Mol. Cryst. Liq. Cryst.* **108**, 359 (1984).
10. Mori T., Kobayashi A., Sasaki Y., Kobayashi H., Saito G., Inokuchi H., *Bull. Chem. Soc. Jpn.* **57**, 627 (1984).
11. Demiralp E., Goddard III W., *J. Phys. Chem.* **98**, 9781 (1994).
12. Sato N., Saito G., Inokuchi H., *Chem. Phys.* **76**, 79 (1983).

Potential surface of rotation-translation coupled systems: Me(NH$_3$)$_6$(PF$_6$)$_2$, Me=Ni,Co.

P. Schiebel*, H.G. Büttner[†], G.J. Kearley[†] M. Prager[‡] and W. Prandl*

*Institut für Kristallographie
Universität Tübingen, Charlottenstr. 33, D-72070 Tübingen
[†]Institut Laue Langevin, Grenoble
[‡] Institut für Festkörperforschung, KFA Jülich, D-52425 Jülich

Abstract. One of the most recent developments in the field of rotational dynamics is the discovery of rotation-translation-coupling (RTC) for molecules in an environment of incompatible symmetry. The combination of incompatible molecular and environmental symmetry usually leads to almost free rotation. Ni- and Co-hexamminehexafluorophosphate thus show large ground state tunnel splittings (0.540 meV (Ni) and 0.542 meV (Co)), as observed by inelastic neutron scattering (INS). The tunnelling spectrum of Ni(NH$_3$)$_6$(PF$_6$)$_2$ is split into three peaks. Under pressure their relative intensities change, but there is almost no shift of the peak positions. In contrast, the tunnelling spectra of Co(NH$_3$)$_6$(PF$_6$)$_2$ shows only one peak, which does not change under pressure.

The hexafluorophosphate compounds are unique among the hexammines, because they do not undergo a phase transition on cooling. From a crystallographic view, both compounds are isostructural (Fm3m). Nevertheless, the proton density distributions, observed by neutron diffraction, although almost circular, show an interesting difference, namely their maxima are rotated by 45 : in the cobalt compound they are on the axes, whereas in the nickel compound they are on the diagonale at room temperature and on the axes at 5K. This is related with a phase shift between the two terms characterizing the potential surface. Obviously, the potential is very sensitive to weak changes of the environment.

Compounds of the family M(NX$_3$)$_6$Y$_2$ with M = Ni, Co; X = H,D and Y = Br, Cl, I, NO$_3$, PF$_6$ are isomorphous. In their high temperature phase they generally form a face centered cubic lattice (space group: Fm3m), where the basic unit is a cube of Y-ions which surround one Ni(NX$_3$)$_6$ octahedron. With the NH$_3$ pyramids on a site with symmetry 4mm it is obvious that they must be orientationally disordered. Whereas the Br-, Cl-, I-, and NO$_3$-compounds undergo phasetransitions on cooling, Ni(NH$_3$)$_6$(PF$_6$)$_2$ and Co(NH$_3$)$_6$(PF$_6$)$_2$ remain cubic down to 5 K.

I INELASTIC NEUTRON SCATTERING

Rotational tunnelling of molecular groups is an extremely sensitive probe to the form and magnitude of the rotational potential. Multiple tunnelling peaks from a single type of group are quite rare, but when they arise, they are attributed to rotor/rotor or rotation-translation-coupling. Ni- and Co-hexamminehexafluorophosphate thus show large ground state tunnel splittings (0.540 meV (Ni) and 0.542 meV (Co)), as observed by inelastic neutron scattering (INS) [1,2]. The tunnelling spectrum of $Ni(NH_3)_6(PF_6)_2$ is split into three peaks (Fig.1). Under pressure their relative intensities change, but there is almost no shift of the peak positions. In contrast, the tunnelling spectra of $Co(NH_3)_6(PF_6)_2$ shows only one peak, which does not change under pressure (Fig.1).

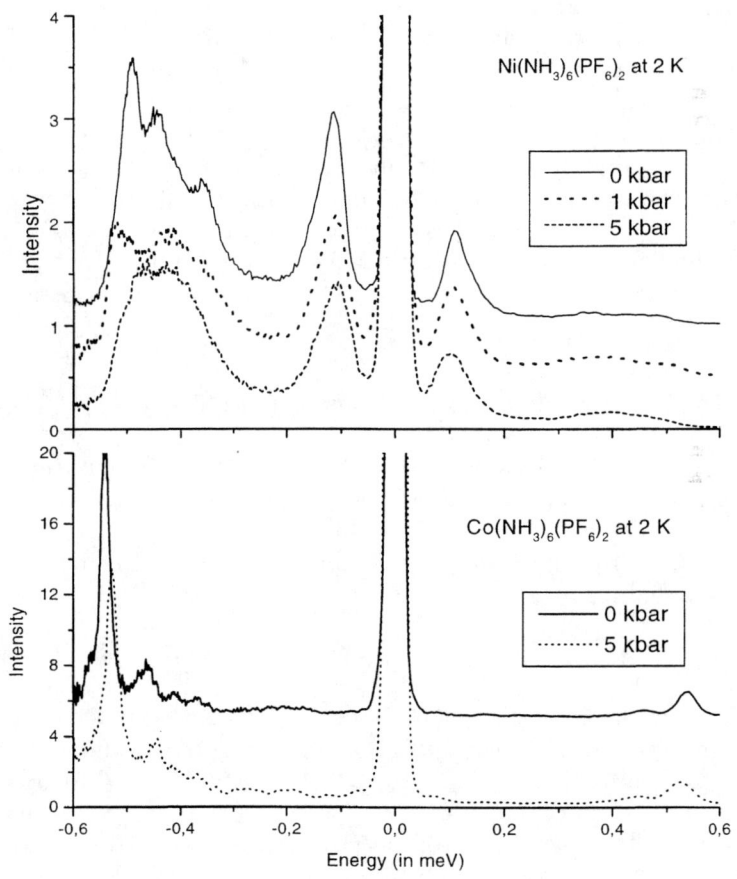

FIGURE 1. Inelastic neutron scattering spectra of NH_3 tunnelling in $Ni(NH_3)_6(PF_6)_2$ (top) and $Co(NH_3)_6(PF_6)_2$ (bottom)

II NEUTRON DIFFRACTION

Diffraction data yield a time averaged density distribution of dynamically disordered molecules. Neutron single crystal diffraction has the best sensitivity to hydrogen and to the effects of thermal motion. Thus, it provides for a direct image of the hydrogen dynamic. For a perfect free uniaxial rotor the protons have equal probability of occupying any position on a circle. The corresponding density distribution is uniform on a ring whose radius corresponds to the radius of rotation. Any deviations from uniform rotation will give rise either to a deformation of the circle or/and a modulation of the density on the circle.

The measured intensity within each Bragg reflection is proportional to the absolut value of the structure factor. We calculate the nuclear density distribution by a combination of conventional crystallographic split-atom density interpolation and Maximum Entropy or Fourier density reconstruction. The crystallographic density interpolation provides model free, correct phases for the measured structure factors.

The proton density obtained in Me(NH$_3$)$_6$(PF$_6$)$_2$ Me=Co,Ni at both temperatures is concentrated on a plane perpendicular to the fourfold crystal axis. Fig. 2 shows cuts through the densities parallel to the face of the unit cube at $z = z_H$ where the maximum proton density occurs. The proton density is found to be nearly circular, with a weak tetragonal contribution superimposed. This is a strong indication of nearly free rotation in these compounds. The weak tetragonal contribution leads to 4 density maxima on the corners of a square. The distance from these maxima to the fourfold axis is larger then the rotation radius of a free NH$_3$ rotor. Therefore the center of mass motion (c.o.m.) has to be included in the dynamical model.

III ROTATION-TRANSLATION COUPLING

We analyse the proton density in terms of an anharmonic orientational potential which couples the rotational motion of the NH$_3$-groups to the translational motion of the center of mass of the three protons [3-5]. This model is obtained by considering the single particle dynamics of an rigid and equilateral ammonia group in a symmetry adapted mean crystal potential

$$V_{Cr}(r,\phi) = \frac{1}{2}Ar^2 + \frac{1}{4}Br^4\cos 4\phi + \frac{1}{4}Cr^4,$$

where A,B,C are three parameters to be determined. (r,ϕ) is the actual position of one hydrogen atom. A given setting of the NH$_3$ group is uniquely defined by the polar coordinates of its c.o.m. (R_c, ϕ_c) and by the angle β associated to the rotation around its center of mass. (r,ϕ) depend on the c.o.m. distance R and on the NH$_3$ setting. The effective molecular potential for one setting is given by

$$V_M(R_c, \phi_c, \beta) = \sum_{p=0}^{2} V_{Cr}(r_p, \phi_p) \tag{1}$$

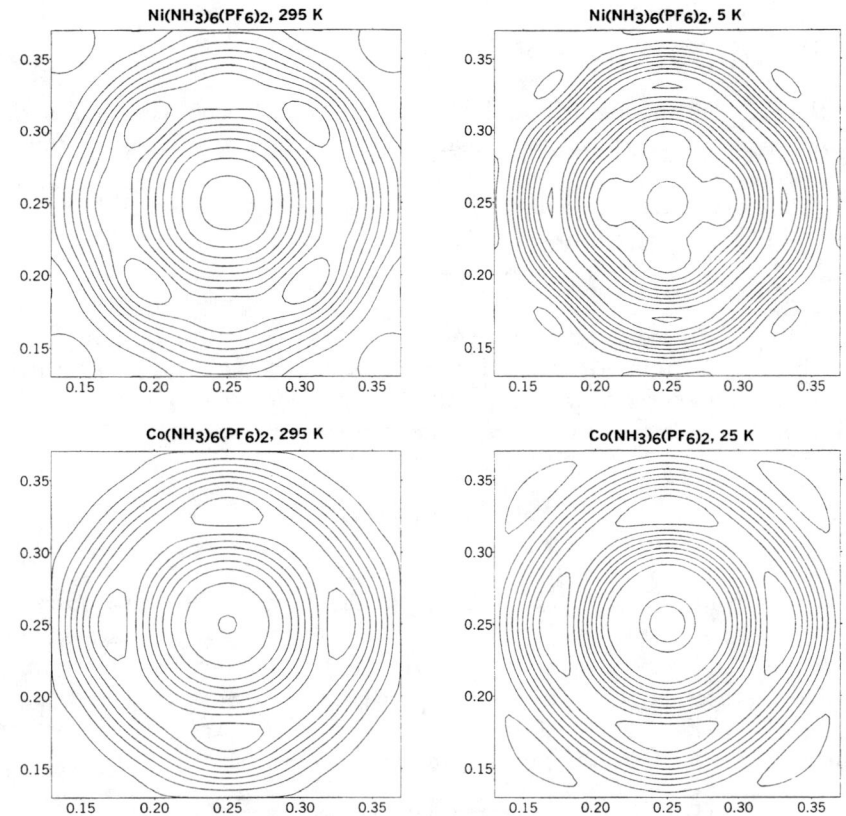

FIGURE 2. Observed nuclear density distribution of the three ammonia hydrogen atoms in the nickel (top) and cobalt (bottom) compound.

The Boltzmann probability for one configuration (R_c, ϕ_c, β) is known to be

$$\rho(R_c, \phi_c, \beta) = Z^{-1} e^{-V_M(R_c,\phi_c,\beta)/kT} \tag{2}$$

and the number density $\rho(x, y)$ observed is a configurational average

$$\rho(x,y) = 3Z^{-1} e^{-V_{Cr}^0(x,y)/kT} \int e^{-(V_{Cr}^1+V_{Cr}^2)/kT} d\gamma. \tag{3}$$

Thus, the mean crystal potentials are determined by least squares analysis of the observed densities.

Low temperature measurements do not allow an interpretation of these phenomena in terms of Boltzmann densities [3,4]. Schrödinger's equation of motion $\mathcal{H}\Psi = E\Psi$ for the problem of a 2D-translational center of mass motion coupled to the uniaxial rotational motion of the molecule is given by

$$\left(\frac{\vec{p}_c}{2M} + BL^2 + V_M(R_c, \phi_c, \alpha)\right)|\Psi> = E|\Psi>, \tag{4}$$

where $\vec{p}_c$ is the momentum operator and $B = \hbar^2/2I$ the rotational constant of the NH_3 molecule. The wavefunctions are expanded into symmetry adapted products of 2D-oscillator and 1-D-rotator wave functions.

Diagonalization of the Hamiltonian leads to the energy eigenvalues and the corresponding eigenfunctions. From the wavefunctions the probability density function

$$\rho(R_c, \phi_c, \alpha) = Z^{-1} \sum exp(-\beta E)|\Psi|^2 \tag{5}$$

of the orientation of the molecule is obtained. The observable proton density distribution is obtained by numerical integration. In addition the energy eigenvalues may be compared to observed tunnel splittings, whereas the intensities of the tunnel lines are obtained by a calculation of transition matrix elements $< \Psi_i | \Psi_j^* >$.

IV DISCUSSION

The potential parameters obtained from the analysis of the proton density distributions are given in Table 1. The rotation of the 4 density maxima observed with $Me(NH_3)_6(PF_6)_2$ is reflected in the change of the sign of the potential parameter B. The derived potentials result in tunnelling levels which are slightly lowered compared to those of a free rotor NH_3 and thus agree with the observation for $Co(NH_3)_6(PF_6)_2$ (Fig.1). However, it is not possible to explain the observed splitting in the tunnelling spectra of the Ni-compound (Fig.1).

TABLE 1. Potential parameters obtained from the observed proton density distributions.

	$Co(NH_3)_6(PF_6)_2$		$Ni(NH_3)_6(PF_6)_2$	
	295 K	25 K	295 K	5 K
A (KÅ^{-2})	-548	-142	-50	-7.4
B (KÅ^{-4})	-63	-19	24	-1.7
C (KÅ^{-4})	1973	332	713	30

REFERENCES

1. Kearley, G. J., Blank, H. and Cockcroft, J. K., *J. Chem. Phys.* **86**, 5987 (1987)
2. Kearley, G. J., Cockcroft, J. K. and Blank, H., *Quantum Aspects of Molecular Motions in Solids* ed. A. Heidemann et al., Springer Proc. in Phys., vol.17, p. 58, Springer Verlag, Berlin, (1987)
3. Schiebel, P. et al., *Z. Phys. B* **81**, 253 (1990).
4. Schiebel, P. et al., *J. Phys. C: Condens. Matter.* **6**, 10989 (1994).
5. Schiebel, P. et al., *Acta Cryst* **A52**, 189 (1996).

Structure and Bonding in *Cis*-enol Systems.

Georg K. H. Madsen

Department of Chemistry
University of Aarhus, DK-8000 Århus C, Denmark

Abstract. The structures and electron density distributions of benzoylacetone (d_{O-O} = 2.502 Å) and nitromalonamide (d_{O-O} = 2.391 Å) have been studied by single crystal X-ray and neutron diffraction at very low temperatures. The neutron experiments have shown that the enol hydrogen atoms are ordered. They are shared between the two oxygen atoms and have large atomic displacement parameters. By combining the neutron and X-ray data experimental electron densities have been obtained. This has shown that the bonding of the enol hydrogen has both an electrostatic and a covalent character.

INTRODUCTION

Cis-enoles constitute an intriguing set of compounds that form strong hydrogen bonds both intra- and intermolecularly.[1,2] The strong hydrogen bonds in *cis*-enol systems involve no charged fragments. Gilli, Bellucci, Ferretti and Bertolasi[3] have proposed the resonance assisted hydrogen bonding (RAHB) model to rationalise the very short O-H···O and N-H···O distances observed in conjugated neutral systems containing hydrogen bonds. Many examples of such conjugated systems have been found among the β-diketone enols, where O-H···O distances could be related to the π delocalization of the O=C-C=C-O-H keto enol group. The essence of the RAHB model for a *cis*-enol fragment is illustrated by scheme a shown in Figure 1. As originally proposed by Gilli et al.[3] the π-electron delocalization introduces partial charges at the oxygens. Consequently, the energy of the system is lowered as the positive hydrogen nucleus moves towards the negative keto oxygen atom. Thus, Gilli's RAHB model can be conceived as a feed back mechanism which maintains zero partial charge on the two opposite oxygens by neutralising the increase in polarization due to resonance with a decrease caused by a shift in the proton position in the hydrogen bond.[4] The RAHB *cis*-enol system is characterized by a large degree of charge delocalization and symmetry of the keto-enol group.

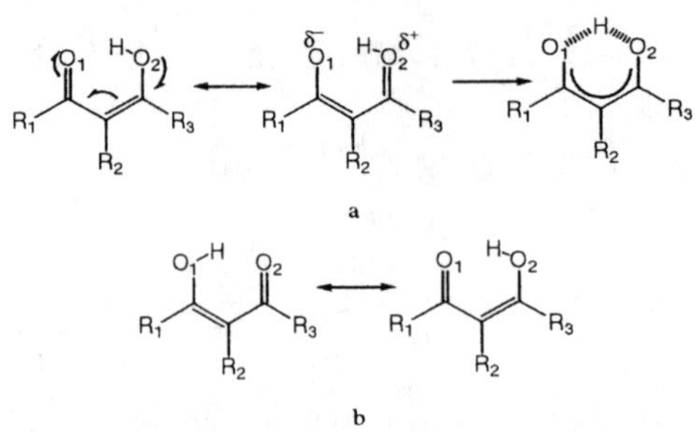

FIGURE 1. a) The resonance assisted hydrogen bonded model. b) The disordered model.

Scheme b in Figure 1 illustrates an alternative explanation for the observation that bond lengths in *cis*-enol systems are intermediate between single and double bonds. If the molecules have statistically disordered keto-enol systems, the hydrogen atoms of the hydrogen bond will be distributed over two positions in the crystal structure, as was the case for the C polymorph of naphthazarin above 110 K.[5] The order/disorder problem corresponds to examining the potential well of the enol hydrogen. Three situations can be imagined: A double well potential where the hydrogen is statistically disordered over two positions; a low barrier potential well where the barrier of protone transfer is lower than the zero point vibrational energy of the hydrogen and, when the O—O distance is short enough, a single well potential.

The symmetry of short hydrogen bonds mean that the bonding of the hydrogen to both the oxygen atoms must be similar. Topological analysis of the electron density[6] has been a resent and succesfull bridge between theoretical chemistry and experimentally determined electron densities. Topological studies of the electron density distributions in three anions have shown that the short hydrogen bonds have covalent character,[7] while an uncharged conjugated system with a short O—O distance has been found to have a localized hydrogen atoms, with a positive Laplacian at the O..H bond critical point.[8]

Better understanding of the structure and nature of short hydrogen bonds is potentially important for understanding enzymatic catalysis. Several enzymatic reactions have been proposed to involve low-barrier hydrogen bonding facilitating enzymatic catalysis.[9] Frey el al.[10] have presented FTIR and NMR evidence for chrymotrypsin and other serine proteases utilize a low-barrier hydrogen bond between His[57] and Asp[102] to facilitate the abstraction of the β-OH proton from Ser[195] in the course of hydrolysis of peptide bonds.

METHODS

The posibility of a disorder made the use of neutron diffraction mandatory to decide the correct structural model.[11] A big advantage of neutron diffraction is furthermore that reliable information on the atomic displacements of the hydrogen can be obtained. Especially for room temperature diffraction studies the observed atomic displacement of the hydrogen atom around its average position is usually so substantial that it is impossible to judge whether the hydrogen atom is distributed over two positions (statistically or residing in a double minimum) or is moving in a shallow potential well. If the thermal energy of the hydrogen atom is sufficiently high, it is conceivable that a double minimum potential may be disguised as dynamic disorder. Thus it is clearly desirable to carry out diffraction studies of hydrogen bonding at the lowest possible temperatures.[12] We have examined the structure and electron density distributions of benzoylacetone (R_1=CH_3 R_2=H, R_2=C_6H_5 , d_{O-O} = 2.502 Å)[13] and nitromalonamide (R_1=R_3=NH_2, R_2=NO_2 ,d_{O-O} = 2.391 Å)[14] by single crystal neutron diffraction at 20K and 15 K respectively. The need for neutron diffraction can be illustrated very clearly by difference Fourier plots calculated without including hydrogen atoms n the model. Figure 2a, the result of an X-ray diffraction study, shows how, as a result of the high formal charge on, and the large atomic displacements of, the enol hydrogen, the thermally smeared electron density is very flat around the position of the enol hydrogen and even might hint, that the hydrogen is disordered over two positions between oxygen atoms O(1) and O(3). However Figure 2b, the result of a neutron diffraction study unequivocally shows the hydrogen position to be ordered.

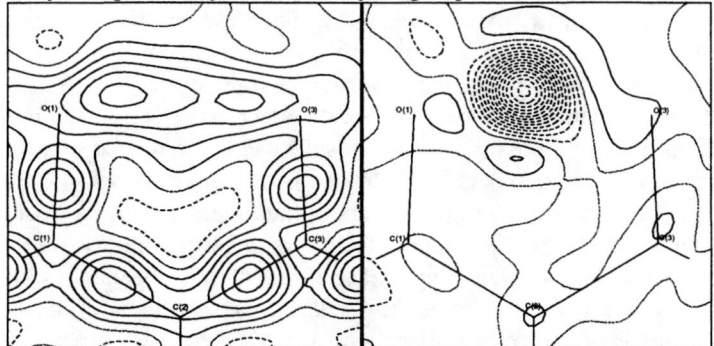

FIGURE 2. Difference Fourier plots for nitromalonamide. a) The X-ray diffraction study, contour levels are 0.1 e/Å^3 b) The neutron diffraction study, contour levels are 0.1 10^{-14} m/Å^3

Once the correct structural model has been found an experimental electron density can be derived by collecting X-ray data at a matching temperature (8K for benzoylacetone and 15K for nitromalonamide). The electron density that is derived from fitting a multipolar model to the measured stucture factors can subsequently be analysed by topological methods to provide a detailed picture of the electron density.

The experimentally determined structures and electron densities can be used as reference points for *ab-initio* calculations that provide supplementary information on the energetics and potential surfaces.

RESULTS

Both benzoylacetone and nitromalonamide have been found to be ordered structures.[13] The enol hydrogen in benzoylacetone is located close to the center of the two oxygen atoms. It exibits a large mean square displacement amplitude (msda) parallel to the O—O vector. Fourth order Gram-Charlier parameters have been refined on the enol hydrogen and have been found to be significant at a 2σ level. These observations have been interpreted as a hydrogen vibrating in a low barrier potential well with sufficient energy to shuttle between the two oxygens. This interpretation has been strengthened by the *ab-initio* calculations on benzoylacetone. It was found that the zero-point vibration energies of the enol hydrogen, estimated by a harmonic extrapolation around the optimised positions, were higher than the barrier of proton transfer. In the molecularly symmetrical nitromalonamide the enol hydrogen is less centered than the enol hydrogen in benzoylacetone, probably because of intermolecular effects. The O—O distance in nitromalonamide lies within the range of a single well potential. The msda parallel to the O—O vector is only about a third of the msda of benzoylacetone and closer to, but higher than, the msda of the hydrogen involved in the short linear charge assisted hydrogen bond in DQNA.[15] *Ab-initio* calculations at the MP2/cc-VDZp and cc-VTZp levels of theory have given a very low barrier height of 0.6 kJ/mol.[16]

The subsequent topological analysis of the experimental electron electron density has provided evidence of a large degree of delocalisation in the enol rings and of the short hydrogen bonds in benzoylacetone and nitromalonamide having covalent character. There are however also large formal charges on oxygen atoms and the enol hydrogen atom, showing that there are both covalent and electrostatic contributions to the stabilisation of the enol hydrogen. Our results can be compared to a similar study of citrinin which has two intramolecular hydrogen bonds of similar O—O distance (d_{O-O} = 2.53 Å and 2.47 Å). At room temperature both tautomers of citrinin exist, while at low temperature only the localised p-quinone form exists. Despite the O—O distances being similar to the one in benzoylacetone the enol rings show distinctly alternating bond lengths and both the hydrogen atoms are localised, closely bound to each their oxygen atom. Correspondingy the bond critical points of the hydrogen atoms each show one strongly negative Laplacian and one strongly positive. With the covalency of the hydrogen bond thus linked to the delocalisation of the system, more than to O—O distance, it could be interesting to consider whether the Laplacian at the bond critical point is really the deciding factor for order/disorder of the enol hydrogen, rather than the O—O distance.

ACKNOWLEDGEMENTS

Finn K. Larsen, Bo B. Iversen, Frank H. Herbstein, Claire Wilson, Garry I. McIntyre and Thomas M. Nymand are all thanked for assistance during both the collection and interpretation of the data.

---

[1] Emsley, J. *Struct. Bond. (Berlin)* **57**, 147-191 (1984).
[2] Hibbert, F. and Emsley, J. *Adv. Phys. Org. Chem.* **26**, 255 – 379 (1990).
[3] a) Gilli, G.; Bellucci, F.; Ferretti, V.; Bertolasi, V. *J. Am. Chem. Soc.* **111**, 1023-1028 (1989). b) Gilli, P.; Ferretti, V.; Bertolasi, V.; Gilli, G. *Advances in Molecular Structure Research*, edited by M. Hargittai and I. Hargittai, Vol. 2, JAI Press, Greenwich, CT, 67-102, 1996.

[4] Bertolasi, V.; Gilli,P.; Ferretti, V.; Gilli, G. *J. Am. Chem. Soc* **113**, 4917-4925 (1991).
[5] Herbstein F. H.; Kapon, M.; Reisner, G. M.; Lehmann M. S.; Kress, R. B.; Wilson R. B.; Shiau, W. I.; Duesler; E. N.; Paul, I. C.; Curtin, D. Y. *Proc. R. Soc. Lond.* **399**, 295-319 (1985).
[6] Bader, R. F. W. *Atoms in molecules. A quantum theory*, Oxford university press, Oxford, 1990.
[7] Flensburg, C.; Larsen, S.; Stewart, R. F. *J. Phys. Chem.* **99**, 10130-10141 (1995). Mallinson, P.R.; Wozniak, K.; Smith, G.T.; McCormack, K.L. *J. Am. Chem. Soc.* **119**, 11502-11509 (1997), Madsen, D.; Flensburg, C.; Larsen, S. *J. Phys. Chem. A* 102, 2177-2188 (1998).
[8] Roversi, P.; Barzaghi, M.; Merati, F. and Destro, R. *Can. J. Chem.* **74**, 1145-1161 (1996)
[9] Cleland, W. W. and Krevoy, M. M. *Science* **264**, 1887-1890 (1994)
[10] Frey, P. A.; Whitt, S. A.; Tobin, J. B. *Science* **264**, 1927-1930 (1994).
[11] see f. ex. Thomas, J. O.; Tellgren, R. and Olovson I. *Acta Cryst.* **B30**, 2540-2549 (1974)
[12] Larsen, F. K. *Acta Cryst.* **B51**, 468-482 (1995).
[13] Madsen, G. K. H.; Iversen, B. B.; Larsen, F. K.; Kapon, M.; Reisner, G. M. and Herbstein, F. H. *J. Am. Chem. Soc.* **120**, 10040-10045 (1998)
[14] Madsen, G. K. H. and Wilson, C. in *Charge, Spin and Momentum Densities and Chemical Reactivity*. Dordrecht: Kluwer Academic. *In Press*
[15] Takusagawa, F. and Koetzle, T. F. *Acta Cryst.* **B35**, 2126-2135 (1979)
[16] Madsen, G. K. H.; Wilson, C.; Nymand, T. M.; McIntyre, G. J.; Larsen, F. K. to be submitted to *J. Phys. Chem. A*.

Simulations of Hydrogen Bonds in Crystals and Their Comparison with Neutron Diffraction Results

Sławomir J. Grabowski

Institute of Chemistry, University in Białystok, Al.J.Piłsudskiego 11,
15-443 Białystok, Poland

Abstract. The influence of the hydrogen bond formation on O-H, N-H and C-H groups have been studied by crystal structure correlation methods on neutron diffraction data taken from Cambridge Structural Database. The correlations between the lengths of the donor groups and the proton...acceptor distances may be supported using the valence sum rule. Hydrogen bonds play a major role in determining molecular packing and conformations of molecules in crystals. Even the weak C-H..X hydrogen bonds which do not influence the donor groups may play a role for the arrangement of molecules in crystals.

Introduction

Hydrogen bonds play an important role in chemistry and biology, they are among the weak electrostatic binding interactions found in nucleic acid structure, protein conformation and supramolecular chemistry. The best method for studying hydrogen bonds in crystals is neutron diffraction since neutrons are scattered by nuclei and permit the location of H as reliably as the other atoms. X-rays are scattered at the atomic shells, and can therefore show H-atoms only poorly. Cambridge Structural Database (CSD) contains results of X-ray and neutron diffraction studies of organics, and organometallic complexes (1). The neutron studies constitute less than 1% of the CSD structure determinations and as a consequence of such situation X-ray results must often be used in the further investigations concerning the geometries of hydrogen bonds. It is therefore of interest to evaluate the H-atom position for the given X-ray hydrogen bonding. The method was suggested by Jeffrey and Lewis (2) where the H-atom is moved along the X-H bond to the position with the standard neutron bond length (O-H = 0.98 Å, N-H = 1.01 Å and C-H = 1.08 Å). However in such treatment the elongation of the X-H bond within X-H..Y system as a result of H-bond formation is not taken into account. Hence, the more refined treatments should be introduced to determine the hydrogen bonding geometry.

The Valence Sum Rule for the Description of Hydrogen Bonds

The strength of the bond and of the intermolecular contact may be measured by its bond valence s. The relationship between the bond valence and the bond length is monotonic and over the small range in which most bonds are found it can be approximated by following equations (3):

$$s_{ij} = \exp[(r_0 - r_{ij})/B] \quad (1)$$

or

$$s_{ij} = (r_{ij}/r_0)^{-1} \quad (2)$$

The shape of the curve described by eqs. 1 and 2 reflects the repulsion between the overlaping electron clouds of the interacting atoms. s_{ij} defines the bond valence of the bond between atoms i and j, r_{ij} is the bond length; r_0, B and N are fitted constants. r_0 is usually the length of the bond of the unit valence. Within the bond valence model atomic valences V_i (being equal to the oxidation states) of atoms are assumed to be shared between the bonds they form:

$$\sum_j s_{ij} = V_i \quad (3)$$

Eq.3 is also called the valence sum rule (3). The aim of the present report is the application of the valence sum rule to the study of O-H..O, N-H..N and C-H..O hydrogen bonds. To investigate the relationships between geometrical parameters within hydrogen bonding systems the Cambridge Structural Database (CSD) was searched for error-free neutron crystal structures, the structural data were restricted to the accurate ones; e.s.d's of the bond lengths ≤ 0.005 Å and $R \leq 5\%$. According to the valence-sum rule (eq.3) we can write for X-H..Y systems:

$$s_{X-H} + s_{H..Y} = 1 \quad (4)$$

where s_{X-H} and $s_{H..Y}$ are the bond valences of X-H and H..Y respectively. Fig.1 presents the correlation between the O-H bond length and the H..O distance for the sample of 33 C=O..H-O-C systems.

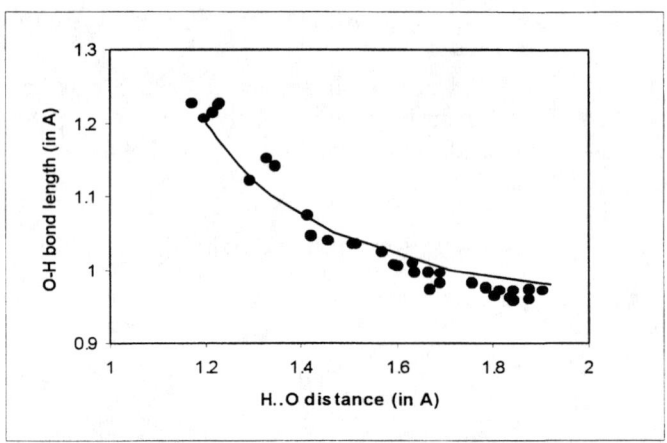

FIGURE 1. Bond length O-H plotted against the H..O distance. Circles correspond to the experimental entries whereas continuous line was obtained from eq.4.

The systems with O-H bond lengths ≥ 0.957 Å (r_0 - value assumed as a length of a single O-H bond - the microwave spectroscopy result for the water molecule) and H..O distances ≤ 2.6 Å were taken into account. The constant value B of eq.1 may be determined from eq.2 applied to fractional O-H bond - $r_{1/2}$:

$$\exp[(r_0 - r_{1/2})/B] = 1/2 \qquad (5)$$

$r_{1/2} = 1.2$ Å is taken from the O..O distance of 2.4 Å for which the proton is centered (4). Than it is possible to find the theoretical correlation between the O-H bond length and the H..O distance (eq.4, fig.1 - the continuous line). The curve obtained from eq.4 agrees with the experimental results in spite of the simplicity of the relation used here. The same attitude as that presented here was applied earlier for the other samples taken from CSD and containing O-H..O systems (4-6). The valence sum rule (eq.3) may be also useful for the analysis of the correlations within N-H..N systems. The data sample selected from CSD and containing N-H..N systems ($R \leq 5\%$, e.s.d's ≤ 0.005 Å, H..N distances are not greater than 2.6 Å) is analyzed according to the procedure applied for O-H..O systems. Fig.2 shows the plot of N-H bond length against H..N distance for experimental data (circles).

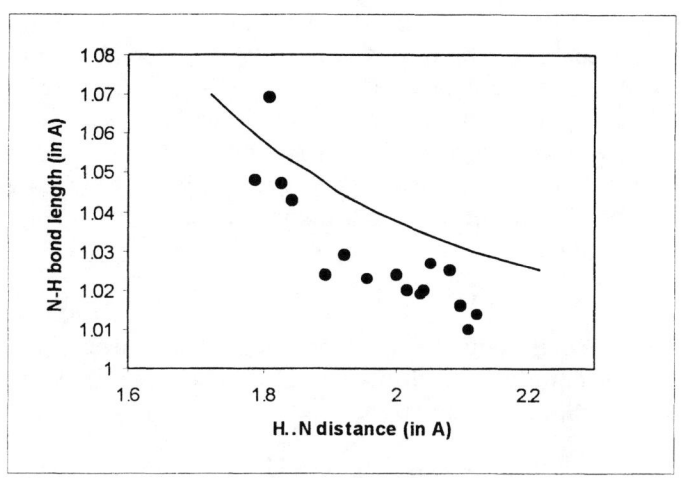

FIGURE 2. Scatter plot of the length of N-H as a function of the H..N distance. Circles show the experimental data taken from CSD and continuous line represents the theoretical curve.

The observed scatter plot is compared with the analytical curve obtained from eqs.4 and 5 ($r_0 = 1.01$ Å and $r_{1/2} = 1.269$ Å) showing that the agreement between experimental data and the theoretical results is not as good as for O-H..O systems. A plot of C-H bond length against hydrogen bond distance H..O is shown in Fig.3.

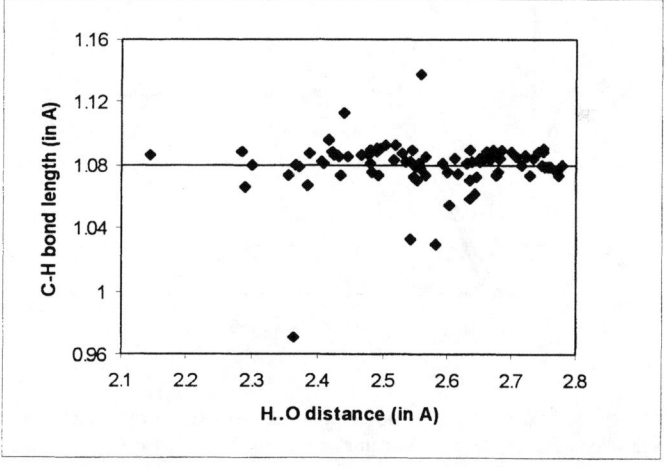

FIGURE 3. Scatter plot of the length of C-H against H..O distance within $C(sp^2)$-H..O systems.

The data sample contains 96 C-H..O systems taken from CSD (R≤5%, e.s.d's ≤0.005 Å, H..O distances not greater than 2.8 Å) for $C(sp^2)$-H donors. The mean value for the C-H bond length is 1.080 Å and the sample standard deviation is 0.017 Å showing that

C-H bond length is not elongated within C-H..O systems, and therefore the valence sum rule was not applied here. The results concerning C-H bond lengths for the sample investigated here are in agreement with the method of 'normalisation' suggested by Jeffrey and Lewis (2); the standard C-H bond length within this attitude is 1.08 Å.

The idea of the valence sum rule may be also refered to C=O bond within C=O..H-O system (6). The atomic valence of oxygen atom shoud be equal to 2 (the oxidation state). Than it is possible to show the following relation:

$$s_{C=O} + s_{H..O} = 2 \qquad (6)$$

where $s_{C=O}$ and $s_{H..O}$ correspond to the appropriate bond valences. It was pointed out that for the neutron diffraction results the correlation between C=O and H..O exists and the experimental results are in agreement with eq.6.

Hydrogen Bond Energy

For many purposes not only the geometry of a hydrogen bond is required but also the determination of its energy. The most general semiempirical treatment for the hydrogen bond energy is the model of Lippincott and Schroeder (7). Fig.4 presents the lowest energies of H-bridges for fixed O..O distances within O-H..O systems (squares).

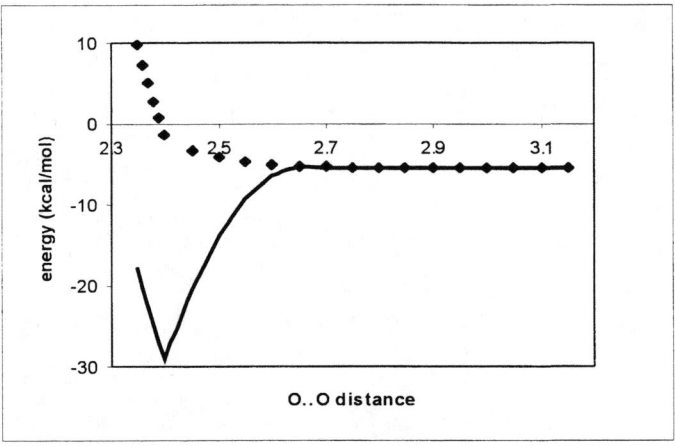

FIGURE 4. The dependence between H-bond energies and O..O distances. Squares represent the Lippincott and Schroeder model and the continuous line its modification.

The curve demonstrates that for the short O..O distances (~2.4 Å) the H-bond energy is possitive and the corresponding systems are not stable. However measurements of pulsed electron-beam mass spectrometry in the gas phase have shown that the largest association entalphies are found for the compounds for which the shortest nearly linear O-H..O bonds are observed (4). Hence the model of Lippincott and Schroeder should be corrected. The appropriate modifications for short O..O distances were proposed (8) and the continuous line at Fig.4 represents the modified H-bond energies. It was pointed out that the continuous line excellently corresponds to the H-bond energies of neutron diffraction O-H..O systems taken from CSD (8).

C-H..C Hydrogen Bonds

Bonds of the types C-H..O and C-H..N have been well-studies in the context of supramolecular chemistry, crystal engineering and molecular recognition. It seems that for C-H..X systems there is no the significant elongation of C-H bond length. However such weak hydrogen bonds influence the arrangement of molecules in crystals. It was pointed out that C-H..O bonds determine crystal packing especially when stronger hydrogen bonding is absent (9). For C-H..C systems a chain of C≡C-H..C≡C-H.. interactions was found in crystalline DL-prop-2-ynylglycine as a cooperative pattern (10). In the Acam polymorphic form of the crystal structure of acetylene (11) studied by neutron diffraction the neighbouring molecules are almost perpendicular (80.4°). The lattice energy calculations were carried out by the use of atom-atom potentials (12) and they showed the values of the angles between neighbouring acetylene molecules in the range 40-50° depending on the type of potentials. For the calculations the interactions within the 729 unit cells were taken into account (i.e. 5832 molecules were considered). After the inclusion of C-H..C potentials derived according to Lippincott and Schroeder model the lattice energies reached a minimum for the angles between neighbouring molecules in the range 85-89°. The similar results regarding the arrangement of acetylene molecules were obtained after the inclusion of C-H..π additional potentials instead of C-H..C. It shows that it is not possible to simulate correctly the crystal structure of acetylene (Acam space group) using the atom-atom potentials unless additional C-H..C or C-H..π interactions are incorporated in the calculations (12).

References

1. Allen,F.H., Kennard,O. and Watson,D.G., Crystallographic Databases: Search and Retrieval of Information from the Cambridge Structural Database. In Structure Correlation, Eds.H.-B.Bürgi and J.D.Dunitz, Weinheim: VCH, Vol.1, pp. 71-110.
2. Jeffrey,G.A. and Lewis,L., Carbohydr.Res. **60**, 179-182 (1978).
3. Brown,I.D., Acta Cryst. **B48**, 553-572 (1992), and references cited therein
4. Gilli,P., Bertolasi,V., Ferretti,V. and Gilli,G., J.Am.Chem.Soc. **116**, 909-915 (1994).
5. Grabowski,S.J., Croat. Chem. Acta **61**, 815-819 (1988).
6. Grabowski,S.J., Tetrahedron **54**, 10153-10160 (1998).
7. Lippincott,E.R. and Schroeder,R., J.Chem.Phys. **23**, 1099-1106 (1955).
8. Grabowski,S.J. and Krygowski,T.M., Tetrahedron **54**, 5683-5694 (1998).
9. Desiraju,G.R., Acc.Chem.Res. **29**, 441-449 (1996), and references cited therein
10. Steiner,T., J.Chem.Soc.,Chem.Commun. 95-96 (1995).
11. McMullan,R.K., Kvick,A. and Popelier,P., Acta Cryst. **B48**, 726-731 (1992)
12. Grabowski,S.J., J.Chem.Res. 534-535 (1996), and references cited therein

LOW-FREQUENCY DYNAMICS

Calculation of Phonon Dispersion Curves by the Direct Method

Krzysztof Parlinski

Institute of Nuclear Physics ul.Radzikowskiego 152, 31-342 Cracow, Poland

Abstract. We report on a direct method which allows, making use of the full crystal symmetry, to calculate the phonon dispersion curves from known Hellmann-Feynman forces found by any *ab initio* program. For non-ionic crystals the direct method reproduces exactly the dispersion curves, provided the range of interaction is confined to the supercell, or it interpolates the dispersion curves between frequencies exactly calculated at special wave vectors, when the interaction range exceeds the supercell size. A windows type software called PHONON, which calculates dispersion curves using the direct method, has been written.

INTRODUCTION

The *ab initio* calculations of phonon frequencies could be carried out within the *linear-response method* or the *direct approach*. Here, we shall concentrate on the direct approach which uses the standard *ab initio* programs based on density functional theory (DTF) and the supercell representation of the crystal. Such programs solves the Kohn-Sham equation within the local density approximation (LDA) or general gradient approximation (GGA) for exchange term, and provides the ground state energy E at $T=0$ for a given atomic configuration. Minimizing E with respect to atomic positions, one may find the crystal structure. Works on *ab initio* phonon dispersion curves started already in 1979 [1]. Early works carried on in the spirit of the direct approach, involved calculating the energy of frozen phonons [2,3].

Traditionally, phonon dispersion relations are calculated by setting up force constants between atoms, constructing the dynamical matrix and diagonalizing it. In the *ab initio* method one derives the force constants by a full quantum mechanical electronic structure calculation for a supercell, and by making a series of small displacements of one atom at a time and calculating so called Hellmann-Feynman (HF) forces exerted on all atoms.

The direct method has already been used to calculate phonon dispersion curves in alkali metals Li, Na, K [4], diamond and graphite [5], chalcopyrites $AgGaSe_2$ [6,7], Si [7], SiO_2 [8], GeS [7,9], triclinic CaP_3 [10], TiC [11] and GaAs [12] crystals. Similar calculations have been carried on for cubic zirconia ZrO_2 [13], and strontium titanate $SrTiO_3$ [14], where soft modes have been found.

BASIC RELATIONSHIPS

Traditionally, the ground state energy E (at $T = 0$) as a function of atomic positions $\mathbf{R}(\mathbf{n}, \mu)$, where $\mathbf{n}$ is the primitive unit cell index and μ is the atomic index, can be expanded over small displacements

$$E(\mathbf{R}(\mathbf{n},\mu),..\mathbf{R}(\mathbf{m},\nu),....) = E_o + \frac{1}{2}\sum_{\mathbf{n},\mu,\mathbf{m},\nu} \Phi(\mathbf{n},\mu;\mathbf{m},\nu)\mathbf{U}(\mathbf{n},\mu)\mathbf{U}(\mathbf{m},\nu) \quad (1)$$

where the *force constant matrix* $\Phi_{i,j}(\mathbf{n},\mu;\mathbf{m},\nu) = \frac{\partial^2 E}{\partial \mathbf{R}_i(\mathbf{n},\mu)\partial \mathbf{R}_j(\mathbf{m},\nu)}\big|_o$ is defined at the extremum configuration $\frac{\partial E}{\partial \mathbf{R}_i(\mathbf{n},\mu)}\big|_o = 0$ at which all first order derivatives of E vanish.

The dynamical matrix is conventionally defined as

$$\mathbf{D}(\mathbf{k};\mu,\nu) = \frac{1}{\sqrt{M_\mu M_\nu}}\sum_{\mathbf{m}} \Phi(0,\mu;\mathbf{m},\nu)\, exp\{-2\pi i \mathbf{k} \cdot [\mathbf{R}(0,\mu) - \mathbf{R}(\mathbf{m},\nu)]\} \quad (2)$$

Here summation $\mathbf{m}$ runs over *all* atoms of the crystal, M_μ, M_ν are atomic masses, and $\mathbf{k}$ is the wave vector. Eigenvalues of the dynamical matrix $\omega^2(\mathbf{k},j)\mathbf{e}(\mathbf{k},j) = \mathbf{D}(\mathbf{k})\mathbf{e}(\mathbf{k},j)$ give the phonon frequencies $\omega^2(\mathbf{k},j)$ and polarization vectors $\mathbf{e}(\mathbf{k},j)$.

Any atomic displacement $\mathbf{U}(\mathbf{m},\nu)$ generates forces $\mathbf{F}(\mathbf{n},\mu) = -\partial E/\partial \mathbf{R}(\mathbf{n},\mu)$ on all other atoms. Using expansion Eq.(1), one finds

$$F_i(\mathbf{n},\mu) = -\sum_{\mathbf{m},\nu} \Phi_{i,j}(\mathbf{n},\mu;\mathbf{m},\nu)U_j(\mathbf{m},\nu) \quad (3)$$

which relates the generated forces with force constant matrices and atomic displacements.

A force constant matrix $\Phi(\mathbf{n},\mu;\mathbf{m},\nu)$ represents a *bond* between two atoms $(\mathbf{n},\mu)$ and $(\mathbf{m},\nu)$. The point symmetry elements b_i, which leave the bond $(\mathbf{n},\mu) - (\mathbf{m},\nu)$ invariant, form a *bond point group* $\hat{\mathcal{B}} = \{b_1, b_2, \ldots b_b\}$. The point symmetry elements b_i occur among the point symmetry elements of the crystal space group $\mathcal{G}$. The force constant matrix is invariant with respect to bond point symmetry elements, i.e. for every b_i, one has $R(b_i)\,\Phi(\mathbf{n},\mu;\mathbf{m},\nu)\,R^T(b_i) = \Phi(\mathbf{n},\mu;\mathbf{m},\nu)$ where $R(b_i)$ is 3×3 transformation matrix representing symmetry element b_i, such as symmetry axis, symmetry plane, inversion, etc. This transformation sets the symmetry of the force constant matrix and reduces the number p of independent parameters. The 3×3 force constant matrix can be equally well written as a (9×1) column matrix, so that $(\Phi)_k = (\Phi)_{i,j}$ and $k = 3(i-1) + j$. Then, every force constant Φ of dimension (9×1) can be decoupled into part $\mathcal{A}$ of dimensions $(9 \times p)$, and part $\mathcal{P}$ of dimensions $(p \times 1)$.

$$\Phi_{(9\times 1)}(\mathbf{n},\mu;\mathbf{m},\nu) = \mathcal{A}_{(9\times p)}(\mathbf{n},\mu;\mathbf{m},\nu) \cdot \mathcal{P}_{(p\times 1)}(\mathbf{n},\mu;\mathbf{m},\nu) \qquad (4)$$

The matrix $\mathcal{A}$ is determined entirely by symmetry, and is independent on potential strength, contrary to $\mathcal{P}$, which collects the values of potential parameters. Every element of $\mathcal{P}$ is an independent parameter, not related with any other by a relation following from crystal symmetry. Decoupling of force constant matrices to symmetry and parameter parts will allow to carry out fitting procedure on independent parameters only, hence on the minimal set of parameters.

SUPERCELL AND FORCE CONSTANTS

Ab initio calculations are always carried out on a *supercell*. The supercell has to be a multiplication of the primitive unit cell. Its lattice constants are the lattice vectors of the crystal. Periodic boundary conditions are always imposed on the supercell. Because of the shape, the parallelopiped of the supercell has its own symmetry described by the *supercell point group* $\hat{\mathcal{S}} = \{s_1, s_2, \ldots s_s\}$. Every symmetry element s_i transforms the supercell (as parallelopiped, not as a crystal) to itself. The supercell acts as a kind of external field imposed on the crystal, and this field may break and lower the crystal symmetry. The point group symmetry of the considered *supercell crystallite* $\hat{\mathcal{H}}$ consists then of point symmetry elements which are common to the supercell point group $\hat{\mathcal{S}}$, and the point group of the studied crystal $\hat{\mathcal{G}}$, i.e.

$$\hat{\mathcal{H}} = \hat{\mathcal{S}} \cap \hat{\mathcal{G}} \qquad (5)$$

Thus, the supercell crystallite space group $\mathcal{H}$ contains common point symmetry elements of $\hat{\mathcal{G}}$ and $\hat{\mathcal{S}}$ and translations and partial translations of the crystal space group $\mathcal{G}$. Consequently, the *supercell crystallite space group*: (i) can be equal to the crystal space group $\mathcal{H} = \mathcal{G}$, or (ii) it can be a subgroup $\mathcal{H} \subset \mathcal{G}$. The following supercell shapes belong to the class (i) and preserve the symmetry of the space group: cubic - $n \times n \times n$, tetragonal - $n \times n \times m$, hexagonal - none, rhombohedral - $n \times n \times n$, orthorhombic - $n \times m \times l$, monoclinic - $n \times m \times l$ and primitive - $n \times m \times l$, where n, m, l are positive integers. In the second case (ii), some equivalent atoms become non-equivalent, hence the site and bond point groups can be reduced to corresponding subgroups. In this case the number of required HF displacements may increase. Consequently, the symmetry of the phonon modes calculated from such a supercell could change.

Consider a supercell and displace an atom $(\mathbf{m}, \nu)$ by $\mathbf{U}(\mathbf{m}, \nu)$. Due to periodic boundary conditions this displacement causes the same displacements of corresponding atoms $(\mathbf{m}+\mathbf{L}, \nu)$ in all images of the supercell. Here, $\mathbf{L} = (L_a, L_b, L_c)$ are the lattice constants of the supercell. Thus, according to Eq.(3), a displacement of a single atom $(\mathbf{m}, \nu)$ in the original supercell generates a net force

$$F_i(\mathbf{n}, \mu) = -\Phi_{i,j}^{(\Sigma)}(\mathbf{n}, \mu; \mathbf{m}, \nu) U_j(\mathbf{m}, \nu) \qquad (6)$$

where the *cummulant force constant* is defined as

$$\Phi_{i,j}^{(\Sigma)}(\mathbf{n},\mu,\mathbf{m},\nu) = \sum_{\mathbf{L}} \Phi_{i,j}(\mathbf{n},\mu;\mathbf{m}+\mathbf{L},\nu) \tag{7}$$

and the summation $\mathbf{L}$ runs over all supercell images. Considering lattice vibrations one should assure that in the dynamical matrix, Eq.(2), all neighbors of a given coordination shell are taken into summation. Hence, the most convenient way when calculating the dynamical matrix, is to locate the center of the supercell at the considered atom $(\mathbf{n},\mu)$. In this manner, one can define an "extended" supercell which has the same size as the original one, but which includes atoms on all its surfaces (edges and corners). Of course, the extended supercell contains more atoms then the conventional one. A special care has to be taken when the diplaced atom $(\mathbf{m},\nu)$ at the original supercell and those at the images $(\mathbf{m}+\mathbf{L},\nu)$ are located at surfaces of the extended supercell and at the same distance from the origin atom $(\mathbf{n},\mu)$. Consider, for example, a displaced atom residing at the corner of the extended supercell. Due to periodic boundary conditions, atoms at all corners of the extended supercell will be displaced by the same amplitude. To take this fact into acount we define a *supercell force constant* as

$$\Phi_{i,j}^{(SC)}(\mathbf{n},\mu;\mathbf{m},\nu) = w_{\mathbf{m}} \Phi_{i,j}^{(\Sigma)}(\mathbf{n},\mu;\mathbf{m}+\mathbf{L},\nu) \tag{8}$$

where the fraction $w_{\mathbf{m}} = 1/n_{\mathbf{m}}$, and $n_{\mathbf{m}}$ is the number of atoms in the set of equivalent atoms on the surfaces of the extended supercell (in the example, number of corners). Simulataneously, Eq.(6) can be written as

$$F_i(\mathbf{n},\mu) = - \sum_{(\mathbf{m},\nu)\in SC} \Phi_{i,j}^{(SC)}(\mathbf{n},\mu;\mathbf{m},\nu) U_j(\mathbf{m},\nu) \tag{9}$$

where the summation $(\mathbf{m},\nu)$ is limited to atoms of the extended supercell including those located at all surfaces. For atomic pairs which are confined to the interior of the extended supercell $w_{\mathbf{m}} = 1$ and the summation over $(\mathbf{m},\nu) \in SC$ reduces to a single term. For atomic pairs which joint the supercell center and surface atom $w_{\mathbf{m}}$ is a fraction. Moreover, such supercell force constant could have a higher symmetry and smaller number of independent parameters than the same conventional force constant, since at the surface $\Phi^{(SC)}$ obeys not only the bond point group symmetry, but also the symmetry provided by the star of all equivalent surface atoms belonging to the same set.

EQUATIONS OF THE DIRECT METHOD

If in Eq.(9) one writes the force constant in the form of a (9×1) column matrix, then it becomes

$$\mathbf{F}(\mathbf{n},\mu) = - \sum_{(\mathbf{m},\nu)\in SC} \mathcal{U}_{(3\times 9)}(\mathbf{m},\nu) \cdot \Phi_{(9\times 1)}^{(SC)}(\mathbf{n},\mu;\mathbf{m},\nu) \tag{10}$$

where the displacement vector $\mathbf{U} = (U_x, U_y, U_z)$ has been converted to (3×9) sparce matrix with elements of $\mathbf{U}$ such that the multiplication rule at Eq.(6) is fulfilled. In analogy to Eq.(4) the supercell force constant can be decoupled into symmetry $\mathcal{A}$ and parameter $\mathcal{P}^{(sc)}$ parts $\Phi^{(SC)}_{(9\times 1)}(\mathbf{n}, \mu; \mathbf{m}\nu) = \mathcal{A}_{(9\times p)}(\mathbf{n}, \mu; \mathbf{m}\nu) \cdot \mathcal{P}^{(SC)}_{(p\times 1)}(\mathbf{n}, \mu; \mathbf{m}\nu)$. Inserting it into Eq.(10), and denoting $\mathbf{C}_{(3\times p)}(\mathbf{n}, \mu; \mathbf{m}, \nu) = -\sum_{(\mathbf{m},\nu)\in SC} \mathcal{U}_{(3\times 9)}(\mathbf{m}, \nu) \cdot \mathcal{A}_{(9\times p)}(\mathbf{n}, \mu; \mathbf{m}\nu)$, Eq.(10) can be simplified to

$$\mathbf{F}(\mathbf{n}, \mu) = \mathbf{C}_{(3\times p)}(\mathbf{n}, \mu; \mathbf{m}, \nu) \cdot \mathcal{P}^{(SC)}_{(p\times 1)}(\mathbf{n}, \mu; \mathbf{m}\nu) \tag{11}$$

or in the global matrix form

$$\mathcal{F} = \mathcal{C} \cdot \mathcal{P}^{(SC)} \tag{12}$$

where $\mathcal{F}$, $\mathcal{C}$ and $\mathcal{P}^{(SC)}$ are matrices of dimensions $(3ns \times 1)$, $(3ns \times p')$ and $(p' \times 1)$. Here, n is the number of atoms in the supercell, s is the number of used displacements for which the HF forces were calculated, and p' is the total number of independent parameters. As a rule $3ns > p'$, and the system Eq.(12) is overdetermined. It may be solved with respect to $\mathcal{P}^{(SC)}$ by the singular value decomposition method [15] applied to the matrix $\mathcal{C}$. The $\mathcal{C}$ matrix is sparce. The direct method produces a solution, which is the best approximation in the least-square sense.

Each set of HF forces should be calculated with a single displaced atom. To obtain the complete information of the values of all force constants it is sufficient to displace each single non-equivalent atom only. Generally, atoms should be displaced along three non-equivalent directions. However, if the site point group is cubic a displacement along single fourfold axis is sufficient. If the site point group is tetragonal one displacement should be taken along fourfold symmetry axis and second one in perpendicular plane to this axis. Lower site point groups require three independent displacements.

SUPERCELL DYNAMICAL MATRIX

In the direct method one defines a *supercell dynamical matrix*, which uses the supercell force constants, Eq.(8),

$$\mathbf{D}^{(SC)}(\mathbf{k}; \mu, \nu) = \frac{1}{\sqrt{M_\mu M_\nu}} \sum_{\mathbf{m}\in SC} \Phi^{(SC)}(0, \mu; \mathbf{m}, \nu) \, exp\{-2\pi i \mathbf{k} \cdot [\mathbf{R}(0, \mu) - \mathbf{R}(\mathbf{m}, \nu)]\} \tag{13}$$

Atom $(0, \mu)$ is always placed at the center of the extended supercell. The summation $\mathbf{m}$ runs over all neighbors of the extended supercell. Summation over images of the supercell is already included in the definition of the supercell force constant, Eqs (7) and (8).

The supercell dynamical matrix, Eq.(13), becomes equal to the conventional dynamical matrix, Eq.(2), $\mathbf{D}^{(SC)}(\mathbf{k};\mu,\nu) = \mathbf{D}(\mathbf{k};\mu,\nu)$ in the following cases:
(i) If the interaction range is confined to the interior of the extended supercell, then the force constants beyond the extended supercell can be neglected. As a result one obtains exactly dipersion curves for *all* wave vectors $\mathbf{k}$.
(ii) If the interaction range speads out beyond the extended supercell, then the two dynamical matrices could become equal at special wave vectors $\mathbf{k}_s$ fulfilling the condition

$$exp\{-2\pi i \mathbf{k}_s \cdot \mathbf{L}\} = 1 \qquad (14)$$

at which the phonon dispersion curves could be calculated exactly. Usually, $\mathbf{k}_s$ correspond to high-symmetry points of the Brillouin zone. Increasing the size of the supercell, one increases the density of the wave vector grid $\mathbf{k}_s$, and interpolates the dispersion curves between the exact points. In this sense the direct method does not impose restrictions on the range of interaction. There exists a windows type program PHONON [16], which uses the direct method to calculate phonons for any crystal from provided HF forces.

REFERENCES

1. Van Camp P.E., Van Doren V.E. and Devreese J.T., *Phys. Rev. Lett.* **42**, 1224 (1979).
2. Kunc K., and Martin R.M., *Phys. Rev. Lett.* **48**, 406 (1982).
3. Yin M.T. and Cohen M.L., *Phys. Rev.* B **26**, 3259 (1982).
4. Frank W., Elsässer C. and Fähle M., *Phys. Rev. Lett.* **74**, 1791 (1995).
5. Kresse G., Furthmüller J. and Hafner J., *Europhys. Lett.* **32**, 729 (1995).
6. Karki B.B., Clark S.J., Warren M.C., Hsueh H.C., Ackland G.J. and Crain J., *J.Phys.: Cond. Matter* **9**, 375 (1997).
7. Ackland G.J., Warren M.C. and Clark S.J., *J.Phys.: Cond. Matter* **9**, 7861 (1997).
8. Karki B.B., Warren M.C., Stixrude L., Ackland G.J. and Crain J., *Phys. Rev.* B **55**, 3465 (1997).
9. Hsueh H.C., Warren M.C., Vass H., Ackland G.J. Clark S.J. and Crain J., *Phys. Rev.* B **53**, 14 806 (1996).
10. Sluiter M.H.F., M.Weinert M. and Kawazoe Y., *Europhysics Lett.* **43**, 183 (1998).
11. Jochym P., Sternik M. and Parlinski K., *Europ. Phys. J.* B submitted.
12. Sternik M., Jochym P. and Parlinski K., *Comput. Matter. Science* in print.
13. Parlinski K., Li Z.Q. and Kawazoe Y., *Phys. Rev. Lett.* **78**, 4063 (1997).
14. Parlinski K., Li Z.Q. and Kawazoe Y., *Phase Transitions* in print.
15. Press W.H., Teukolsky S.A., Vetterling W.T. and Flannery B.P., *Numerical Recepies*, Cambridge: University Press, Cambridge, 1992, pp.670.
16. Parlinski K., Software PHONON (1998).

Phonons in Chalcopyrite Compounds

P. Derollez*, A. Laamyem*, R. Fouret*
B. Hennion† and J. Gonzalez‡

*Laboratoire de Dynamique et Structure des Matériaux Moléculaires
Université de Lille I, 59655 Villeneuve d'Ascq Cedex, France
†Laboratoire Léon Brillouin, Centre d'Etudes Nucléaires de Saclay
91191 Gif sur Yvette, France
‡Centro de Estudios de Semiconductores, Facultad de Ciencias
Universidad de Los Andes, Merida 5101, Venezuela

Abstract. The phonon dispersion curves along the [100] and [001] directions of $CuInSe_2$ and $AgGaSe_2$ have been measured by inelastic neutron scattering. They are analyzed with different rigid-ion models: Born-von Karman and valence force field models. The calculated dispersion curves are in good agreement with experiments.

INTRODUCTION

The structure of chalcopyrite compounds with ABC_2 formula has a tetragonal symmetry with space group $I\bar{4}2d$. It can be considered as being derived from the zinc-blende structure by doubling the conventional unit cell along the c axis. In most chalcopyrite materials, the ratio $\eta = c \div 2a$ is not equal to 1. In addition, chalcopyrite compounds have two kinds of cations and hence two bond lengths: R_{A-C} and R_{B-C}. Up to now, only dynamics at the centre of the Brillouin zone have been investigated [1–3]. In order to obtain a better knowledge of the restoring forces, we have measured the phonon dispersion curves in $CuInSe_2$ and $AgGaSe_2$. The results are analyzed with different rigid-ion models.
In this paper, we report more specially on the results of $CuInSe_2$ (CIS). A previous study on $AgGaSe_2$ [4] will allow comparisons between both compounds. The lattice parameters of CIS at 300 K are a=5.782 Å and c=11.620 Å [5]. The difference between R_{Cu-Se} and R_{In-Se} can be characterized by a dimensionless coordinate u=0.235. The primitive cell include eight atoms whose fractional coordinates can easily be obtained from Table 1 of [4].

EXPERIMENTAL RESULTS

Inelastic neutron scattering experiments have been carried out at room temperature and atmospheric pressure at the Orphée reactor at the Laboratory Léon Brillouin on the triple axis spectrometer 1T. The phonon branches have been measured in the whole energy range in the [001] and [100] directions. The data have been systematically corrected from instrumental resolution effect. There are eight atoms in the primitive cell, therefore for a given direction of the wave vector there are twenty four phonon branches. For wave vectors in the [001] direction, the degeneracies of modes reduce this number to 17. The phonon branches measured in $CuInSe_2$ are shown in Figure 1. Excepting the region near 2 THz, their shapes are similar to those obtained in $AgGaSe_2$ [4].

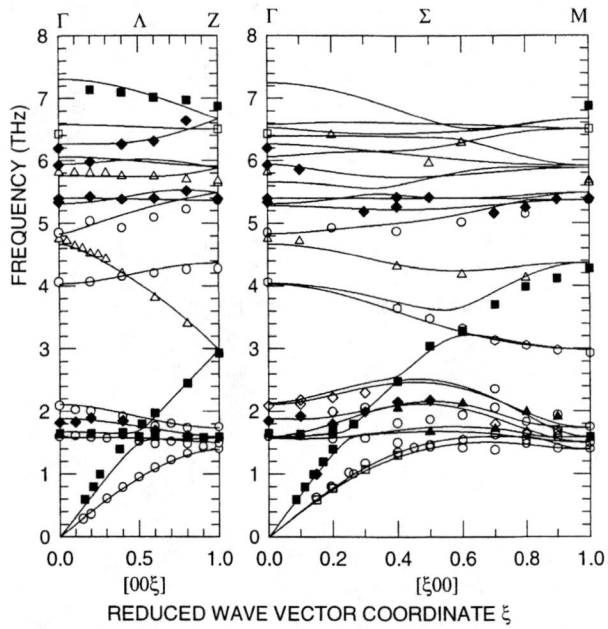

FIGURE 1. Dispersion curves of phonons along the [001] and [100] directions in $CuInSe_2$. The points correspond to the neutron measurements, the full line to the Born-von Karman model.

DYNAMICAL MODELS AND DISCUSSION

A Born-von Karman (BvK) model including short range forces and electrostatic forces was used. The Coulomb forces were calculated following the Ewald summation method. Short range forces have been assumed to be central forces an their contribution to the dynamical matrix is thus characterized by the longitudinal (denoted L) and transverse (denoted T) force constants. We have taken into account short range force constants up to an extend of 5 Å which include interactions between first- and second-neighbour ions. To avoid the multiplicity of adjustable parameters, the same values were used for equivalent pairs with slightly different distances as seen in Table 1. From the LO-TO splitting in CuInSe$_2$, Tanino [1] has

TABLE 1. Equilibrium distances and atomic force constants for the considered interactions in CuInSe$_2$.

Interaction	Shell radius Å	L(Nm^{-1})	T(Nm^{-1})
Cu1-Se1	2.46	54.8	-7.0
	4.77-4.84-4.87	-0.3	1.9
In1-Se1	2.56	120.7	-15.1
	4.72-4.79-4.82	5.3	0.5
Cu1-Cu2	4.10	7.0	0.6
Cu1-In1	4.10	0.2	-2.1
Cu1-In2	4.09	0.2	-2.1
In1-In2	4.10	-7.9	1.0
Se1-Se2	4.04	2.4	2.1
Se1-Se4	3.97	2.4	2.1
Se1-Se2	4.16	2.4	2.1
Se1-Se4	4.21	2.4	2.1

determined the following effective charges: $Z_{Cu} = 0.475$, $Z_{In} = 1.255$ and $Z_{Se} = -0.865$. These values were kept fixed in the refinements since no significant improvement was obtained when trying to refine them. In the refinements with the Unisoft program package [6] linked to a least-squares fit routine, 16 force constants were adjusted, leading to a reliability factor of 1.9%. The final values of the parameters are given in Table 1. The standard deviations for the preponderant force constants are about 3%, the other parameters are much smaller because of the interatomic distances and less well defined (standard deviations less than 20%). The phonon dispersion curves calculated for CuInSe$_2$ are shown in Figure 1. Generally, they describe closely the experimental branches.

The predominant stretching constants are close to the corresponding constants determined by Fouret [4] in AgGaSe$_2$: $L_{Ag-Se}=55.5$ and $L_{Ga-Se}=117.1$ (in Nm^{-1} unit), in a ratio 2:1 for each stretching interaction. On the other hand, the transverse force constants differ notably when passing from CuInSe$_2$ to AgGaSe$_2$ ($T_{Ag-Se}=-9.7$ and $T_{Ga-Se}=-5.1$). This difference originates mainly from the low frequency optical modes which are very sensitive to these transverse constants. In CuInSe$_2$, their frequencies are included between 1.6 and 2.1 THz while the frequency range is larger (0.8–2.4 THz) in AgGaSe$_2$ [4].

The valence force-field model which constitutes a more realistic model in such tetrahedrally coordinated crystals was also used. The short range interatomic forces lead to 2 stretching and 5 bending constants [4] which did not provide as good results as the previous model (reliability factor of 7.9%). Although the stretching constant

α_{Cu-Se} and α_{In-Se} values giving the general shape of the dispersion curves are close than those derived from the BvK model, the bending forces are not sufficient to describe the details of the different branches. The shell model which accounts for electronic polarization leads to a reliability factor slighly better than those of the BvK model (reliability factor of 1.5%) but 36 adjustable parameters were introduced in the refinements.

The following concluding points have been reached as a result of the dynamical studies of chalcopyrite compounds:
(i) Despite of its simplicity, the BvK model gives a realistic description of the phonon dispersion curves of $CuInSe_2$ and $AgGaSe_2$. This can lead to more fruitful conclusions when comparing the force constants.
(ii) The obtained results provide a strong experimental base in view of the validation of *ab-initio* calculations. Ackland *et al.* [7] have discussed the treatment of errors in *ab initio* forces and showed the importance of correcting for errors in the case of $AgGaSe_2$. They have obtained a general description of the experimental dispersion curves. The calculated frequencies at Γ point are in good agreement with experiment (within 0.3 THz). However, some calculated frequencies at the Brillouin zone boundary show discrepancies up to 1 THz with the experimental data, specially in the low frequency range.

ACKNOWLEDGMENTS

The authors wish to thank Dr G. Eckold who provides the program package UNISOFT. One of us (J.G.) acknowledges the CEFI-PCP (France) and the CONICIT-Materials (Venezuela). This work was performed under the frame work of E.E.C. Contract n°CI1-CT94-0031. The Laboratoire de Dynamique et Structure des Matériaux Moléculaires is Unité associée au CNRS n°801.

REFERENCES

1. Tanino H., Maeda T., Fujikake H., Nakanishi H., Endo S. and Irie T., *Phys. Rev. B* **45**, 13323 (1992).
2. Syrbu N.N., Bogdanash M., Tezlevan V.E. and Mushcutariu I., *Physica B* **229**, 199 (1997).
3. Camassel J., Artus L. and Pascual J., *Phys. Rev. B* **41**, 5717 (1990).
4. Fouret R., Derollez P., Laamyem A., Hennion B. and Gonzalez J., *J. Phys.: Condens. Matter* **9**, 6579 (1997).
5. Jaffe J.E. and Zunger A., *Phys. Rev. B* **29**, 1882 (1984).
6. Eckold G., Unisoft: a program package for lattice dynamical calculations (1992).
7. Ackland G.J., Warren M.C. and Clark S.J., *J. Phys.: Condens. Matter* **9**, 7861 (1997).

Frequency Dependent Specific Heat of Amorphous Silica: A Molecular Dynamics Computer Simulation

Peter Scheidler, Walter Kob, Jürgen Horbach, and Kurt Binder

Institute of Physics, Johannes Gutenberg University, Staudinger Weg 7, D-55099 Mainz, Germany

Abstract. We use molecular dynamics computer simulations to calculate the frequency dependence of the specific heat of a SiO_2 melt. The ions interact with the BKS potential and the simulations are done in the NVE ensemble. We find that the frequency dependence of the specific heat shows qualitatively the same behavior as the one of structural quantities, in that at high frequencies a microscopic peak is observed and at low frequencies an α-peak, the location of which quickly moves to lower frequencies when the temperature is decreased.

INTRODUCTION

The dynamics of supercooled liquids can be studied by many different techniques, such as light and neutron scattering, dielectric measurements, NMR, or frequency dependent specific heat measurements, to name a few [1]. In order to arrive at a better understanding of these systems also various types of computer simulations have been used to supplement the experimental data. However, essentially all of these simulations have focussed on the investigation of *static* properties or have studied the time dependence of *structural* quantities, like the mean squared displacement of a tagged particle or the decay of the intermediate scattering function. What these simulations have not addressed so far, apart from a noticeable study of Grest and Nagel [2], is the time dependence of thermodynamic quantities, like the specific heat. The reason for the lack of simulations in this direction is that the accurate determination of this quantity in a simulation is very demanding in computer resources because of its collective nature. This fact is of course very regrettable since one of the simplest ways to determine the glass transition temperature in a real experiment is to measure the (static) specific heat. Using ac techniques it is today also possible to measure the *frequency dependent* specific heat, $c(\nu)$, and thus to gain more insight into this observable [3]. What is so far not possible in real experiments is to measure $c(\nu)$ at frequencies higher than 1MHz, and thus the influence of the microscopic dynamics, which is in the THz range, cannot be investigated. For computer simulations it is, however, no problem to study $c(\nu)$ also at these high frequencies and in this paper we report the outcome of such an investigation for the strong glass former silica.

MODEL AND DETAILS OF THE SIMULATION

The silica model we use is the one proposed by van Beest et al. [4]. In this model the interaction $\phi(r_{ij})$ between two particles i and j a distance r_{ij} apart is given by a two body potential of the form

$$\phi(r_{ij}) = \frac{q_i q_j e^2}{r_{ij}} + A_{ij}\exp(-B_{ij}r_{ij}) - \frac{C_{ij}}{r_{ij}^6} \quad . \tag{1}$$

The values of the partial charges q_i and the constants A_{ij}, B_{ij} can be found in Ref. [4]. Since the quantity we want to investigate, $c(\nu)$, is a collective one, it is necessary to average it over many independent realizations. Thus the system sizes we used are rather small, 336 ions, despite the fact that the dynamics of such a small system will show appreciable finite size effects [5]. However, exploratory runs with larger systems showed that these effects do not change the results substantially. The simulations were done at constant volume using a box size of 16.8 Å, thus at a density of 2.36g/cm^3, close to the experimental value of the density, which is at 2.2g/cm^3. The equations of motion have been integrated with the velocity form of the Verlet algorithm with a time step of 1.6 fs. The temperatures investigated were 6100 K, 4700 K, 4000 K, 3580 K, 3250 K, and 3000 K. At all temperatures the system was first equilibrated for a time which is significantly longer than the typical α-relaxation time of the system at this temperature.

RESULTS

In real experiments the frequency dependent specific heat is usually measured in the NPT ensemble. Although algorithms exist with which the *static equilibrium* properties of a system can be measured in a simulation in this ensemble, these algorithms introduce an artificial dynamics of the particles and are therefore not suited to investigate the dynamical properties of the system in this ensemble. Hence we calculated the frequency dependent specific heat in the microcanonical ensemble. Whereas in the NPT ensemble the specific heat is related to the fluctuations of the enthalpy, in the NVE ensemble it is related to the fluctuations of the kinetic energy [2,6]. It can be shown that in this ensemble the specific heat at frequency ν is given by

$$c(\nu) = \frac{k_B}{2/3 - K(t=0) - i2\pi\nu \int_0^\infty dt \exp(i2\pi\nu t) K(t)} \quad , \tag{2}$$

where $K(t)$ is the autocorrelation function of the kinetic energy E_{kin} and is defined as

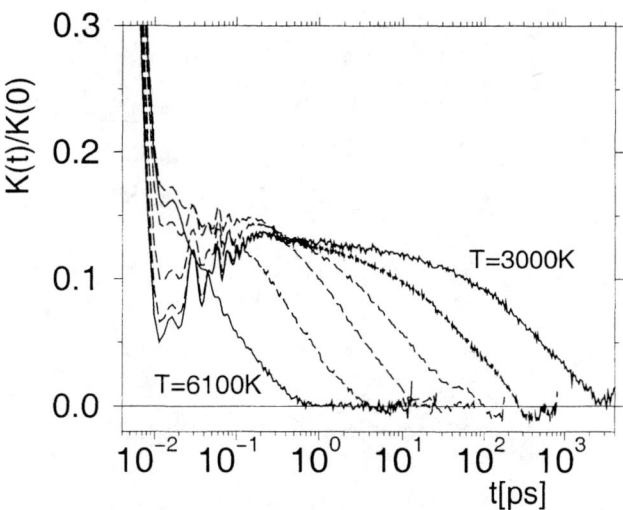

FIGURE 1. Time dependence of the autocorrelation function of the kinetic energy for all temperatures investigated.

$$K(t) = \frac{N}{\bar{E}_{\text{kin}}^2} \left[(E_{\text{kin}}(t) - \bar{E}_{\text{kin}})(E_{\text{kin}}(0) - \bar{E}_{\text{kin}}) \right] \quad . \tag{3}$$

Here $\bar{E}_{\text{kin}}$ is the mean kinetic energy and N is the total number of ions. The derivation of Eq. (2) can be found in Ref. [6].

From Eq. (2) we see that the relevant quantity is the autocorrelation function $K(t)$. The time dependence of this quantity, normalized by its value at time $t = 0$, is shown in Fig. 1 for all the temperatures we investigated.

From this figure we recognize that for high temperatures $K(t)/K(0)$ decays very quickly to a value around 0.13 and then goes to zero like a stretched exponential. With decreasing temperature the function shows a plateau at intermediate times, the length of which increases quickly with decreasing temperature. Such a time and temperature dependence is very similar to the one found for the relaxation behavior of structural quantities, such as the intermediate scattering function [7]. Apart from these features the curves for the lowest temperatures show also a local minimum at around 0.02 ps, the depth of which increases with decreasing temperatures. The existence of this dip, as well as the observed high frequency oscillations, can be understood by realizing that within the harmonic approximation, which will be valid at even lower temperatures, the correlator is closely related to the autocorrelation function of the velocity, which is well known to show such a dip.

Using Eq. (2) we calculated from $K(t)$ the frequency dependent specific heat $c(\nu)$. The real and imaginary part of this quantity are shown in Fig. 2 for all temperatures investigated. Let us first discuss $c'(\nu)$: For very high frequencies we expect this function to go to the ideal gas value of 1.5, since the configurational degrees of freedom are not able to take up energy at such high frequencies. With decreasing

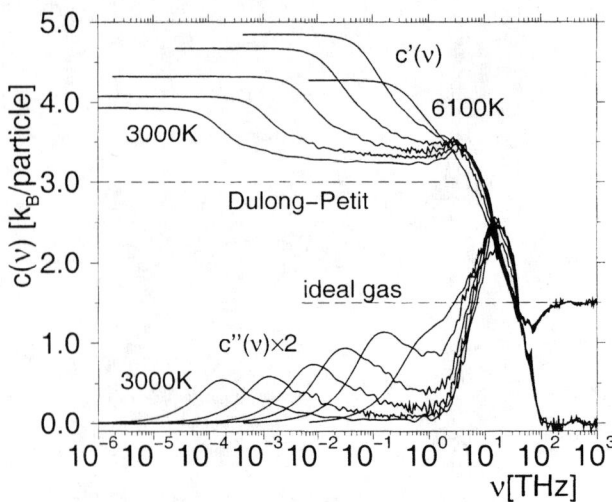

FIGURE 2. Frequency dependence of the real and imaginary part (times 2.0) of the specific heat for all temperatures investigated. The upper and lower dashed lines are the values of $c'(\nu)$ for the harmonic solid and the ideal gas, respectively.

ν the function shows a fast increase in the frequency range which corresponds to the microscopic vibrations. For low temperatures this regime is followed by a plateau the height of which corresponds to the static specific heat of the system *if no relaxation would take place*, i.e. to the specific heat of the vibrational degrees of freedom. Since, however, on the time scale of the α-relaxation time the system relaxes, $c'(\nu)$ shows at the corresponding frequencies a further upward step. This feature is related to the fact that for very long times, or small frequencies, those configurational degrees of freedom which are not of vibrational type are relaxing and thus can take up energy. At even smaller frequencies the curves then show a plateau, the height of which is the static specific heat of the system. We see that with decreasing temperature this value is decreasing but is always significantly above the harmonic value given by the Dulong-Petit value of 3.0, since the relaxing configurational degrees of freedom give rise to an enhancement of the static specific heat.

We also note that the height of the first step in $c'(\nu)$ [coming from the low frequency side] is the configurational part of the specific heat. We see that at the lowest temperature this height is rather small, $0.7 k_B$ per particle, in agreement with the experimental observation for strong glass-formers.

All these features can also be seen well in the imaginary part of $c(\nu)$. At high frequencies we have a microscopic peak which corresponds to the vibrational degrees of freedom. The location and the height of this peak is essentially independent of temperature. At intermediate and low temperatures a second peak is seen at low frequencies, the so-called α-peak. Its position depends strongly on temperature in agreement with the observation that the α-relaxation time of structural quantities increases quickly with decreasing temperatures [7]. From the location of this peak we can read off $\nu_{\max}$, the frequency scale of the relaxation of the specific heat. As it will be shown elsewhere [6], the product of $\nu_{\max}$ with $\tau(T)$, the α-relaxation time of the intermediate scattering function, is essentially independent of the temperature, thus showing the intimate connection between the frequency dependent specific heat and the structural relaxation, in agreement with the prediction of Götze and Latz [8].

To summarize we can say that we have presented the results of a large scale molecular dynamics computer simulation of a realistic model of viscous silica to investigate the frequency dependence of the specific heat. In the frequency regime which is accessible also to experiments our results are in qualitative agreement with the experimental data [3]. At higher frequencies we see the influence of the vibrational degrees of freedom on $c(\nu)$. Since no experimental data for $c(\nu)$ is available for silica we are not able to compare the results of the present simulations with reality. However, in a previous investigation we have shown that the present model gives very good quantitative agreement of the *static* specific heat with the one of real silica [9] and thus it is not unreasonable to assume that the results of the simulation on the dynamic quantity is reliable also.

ACKNOWLEDGMENTS

We thank U. Fotheringham for suggesting this work and A. Latz for many helpful discussions. Part of this work was supported by Schott Glaswerke, by SFB 262/D1 of the Deutsche Forschungsgemeinschaft, and BMBF Project 03 N 8008 C.

REFERENCES

1. See, e.g., articles in the proceeding of the *Third International Discussion Meeting on Relaxation in Complex Systems*, J. Non-Cryst. Solids **235-237** (1998).
2. G. S. Grest and S. R. Nagel, J. Phys. Chem. **91**, 4916 (1987).
3. N. O. Birge and S. R. Nagel, Phys. Rev. Lett. **54**, 2674 (1985); Y. H. Jeong and I. K. Moon, Phys. Rev. B **52**, 6381 (1995); T. Christensen, N. B. Olsen, J. Non-Cryst. Solids **235-237**, 296 (1998) and references in these papers.
4. B. W. H. van Beest, G. J. Kramer, and R. A. van Santen, Phys. Rev. Lett. **64**, 1955 (1990).
5. J. Horbach, W. Kob, K. Binder, and C. A. Angell, Phys. Rev. E **54**, R5897 (1996).
6. P. Scheidler, W. Kob, J. Horbach, and K. Binder, to be published; P. Scheidler, Diploma thesis (University of Mainz, 1999); J. K. Nielsen and J. C. Dyre, Phys. Rev. B **54**, 15754 (1996).
7. J. Horbach, W. Kob, and K. Binder, Phil. Mag. B **77**, 297 (1998); J. Horbach and W. Kob, preprint cond-mat/9901067.
8. W. Götze and A. Latz, J. Phys.: Condens. Matter **1**, 4169 (1989).
9. J. Horbach, W. Kob and K. Binder, J. Phys. Chem. B (in press); see also preprint cond-mat/9809229.

The Boson Peak in Amorphous Silica: Results from Molecular Dynamics Computer Simulations

Jürgen Horbach, Walter Kob, and Kurt Binder

Institute of Physics, Johannes Gutenberg University, Staudinger Weg 7, D-55099 Mainz, Germany

Abstract. We investigate a prominent vibrational feature in amorphous silica, the so-called boson peak, by means of molecular dynamics computer simulations. The dynamic structure factor $S(q,\nu)$ in the liquid, as well as in the glass state, scales roughly with temperature, in agreement with the harmonic approximation. By varying the size of the system and the masses of silicon and oxygen we show that the excitations giving rise to the boson peak are due to the coupling to transverse acoustic modes.

I INTRODUCTION

In the last few years various scattering techniques, such as neutron, Raman and X-ray scattering, have been used to investigate the so-called boson peak, a vibrational feature, which is found in the frequency spectra of many, typically strong, glass formers at a frequency of about 1 THz [1]. In this context various mechanisms giving rise to this peak have been proposed, such as certain localized vibrational modes or scattering of acoustic waves, and also simple models have been developed that produce an excess over the Debye behavior in the density of states [2].

Especially in the case of silica molecular dynamics computer simulations have recently been used in order to gain insight into the nature of the boson peak [3-5]. Despite the limitations of these simulations, such as the small system size (of the order of 10^3–10^4 particles) and high cooling rates (of the order of 10^{12} K/s), they are very useful because they include in principle the full microscopic information in form of the particle trajectories. Most of the recent computer simulation studies have investigated the boson peak within the harmonic approximation in that the eigenvalues and eigenvectors of the dynamical matrix have been calculated [3,4]. In contrast to this method we use the full microscopic information to determine quantities like the dynamic structure factor $S(q,\nu)$ directly from the particle coordinates. Thus, we are not restricted to the harmonic approximation and we are able to compare the dynamics of our silica model in the liquid state with the dynamics in the glass state. Moreover, by varying parameters like the size of the system and the mass of the particles we gain information on the character of the boson peak excitations.

II DETAILS OF THE SIMULATION

The silica model we use for our simulation is the one proposed by van Beest et al. [6] which is given by

$$u(r_{ij}) = \frac{q_i q_j e^2}{r_{ij}} + A_{ij} \exp(-B_{ij} r_{ij}) - \frac{C_{ij}}{r_{ij}^6} \quad \text{with} \quad i,j \in \{\text{Si}, \text{O}\}. \quad (1)$$

The values of the partial charges q_i and the constants A_{ij}, B_{ij} and C_{ij} can be found in the original publication. The simulations were done at constant volume keeping the density fixed at 2.37 g/cm^3. Our simulation box contains 8016 particles with a box length of 48.37 Å. We investigate the equilibrium dynamics of the liquid state as well as the glass state. The lowest temperature for which we were able to fully equilibrate our system was 2750 K. At this temperature we integrated the equations of motion over 13 million time steps of 1.6 fs, thus over a time span of about 21 ns. The glass state was produced by starting from two equilibrium configurations at $T = 2900$ K and cooling them to the temperatures $T = 1670$ K, 1050 K and 300 K with a cooling rate of $1.8 \cdot 10^{12}$ K/s. The details of how we calculated the time Fourier transformations can be found elsewhere [7].

III RESULTS

We investigate the high frequency dynamics of silica by means of the dynamic structure factor

$$S(q, \nu) = N^{-1} \int_{-\infty}^{\infty} dt \exp(i2\pi\nu t) \sum_{kl} \langle \exp(i\mathbf{q} \cdot [\mathbf{r}_k(t) - \mathbf{r}_l(0)]) \rangle \quad , \quad (2)$$

and its self part $S_s(q, \nu)$ which can be extracted from Eq. (2) by taking into account only the terms with $k = l$. In the following we will consider only $S(q, \nu)$ for the oxygen–oxygen correlations because the oxygen–silicon and the silicon–silicon correlations behave similarly with respect to the features which are discussed below.

As we have reported elsewhere for the liquid state at the temperature $T = 2900$ K [5], apart from optical modes with frequencies $\nu > 20$ THZ, two types of excitations are visible in the dynamic structure factor for $q > 0.23$ Å^{-1}. The first one corresponds to the boson peak which is located, essentially independent of q, around 1.8 THz. The second one corresponds to dispersive longitudinal acoustic modes. Note that the latter are not like longitudinal acoustic excitations in harmonic crystals because, due to the disorder, they cannot be described as plane waves.

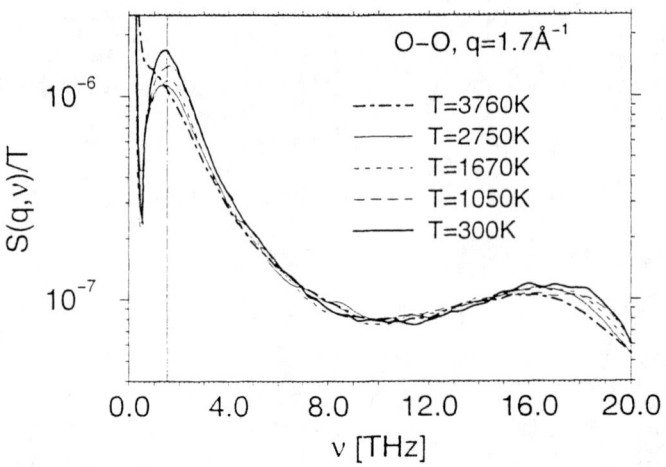

FIGURE 1. $S(q,\nu)/T$ as a function of frequency for different temperatures.

Having found the aforementioned two features at $T = 2900$ K we will now look at the temperature dependence of $S(q,\nu)$, which is shown in Fig. 1, by plotting $S(q,\nu)/T$ versus ν at $q = 1.7$ Å^{-1} in the frequency range below 20 THz. From this figure we can conclude that the dynamic structure factor scales roughly with temperature which is expected if the harmonic approximation is valid. Moreover we can clearly identify for all temperatures two peaks: the boson peak located at 1.75 THz (vertical line) and a peak which corresponds to the longitudinal acoustic modes located around 17 THz. Even at $T = 3760$ K the excitations giving rise to a boson peak at lower temperatures are at least partially present in that a shoulder can be recognized in the frequency region of the boson peak.

In order to get some insight into the properties of the vibrational modes of our silica system at frequencies around 1 THz we varied the size of the system at a fixed mass density of 2.37 g/cm^3. Fig. 2 shows the self part of the dynamic structure factor for $N = 336$, 1002 and 8016 particles at the temperature $T = 3760$ K and the three q-values 0.37 Å^{-1}, 1.7 Å^{-1} and 4.75 Å^{-1}. Whereas the curves for the different system sizes coincide for frequencies that are larger than a weakly N dependent frequency $\nu_{cut}(N)$, for $\nu < \nu_{cut}(N)$ the amplitude of $S_s(q,\nu)$ decreases with decreasing N. Note that $\nu_{cut}(N)$ is essentially independent of the wave-vector q. We read off $\nu_{cut} \approx 1.7$ THz for $N = 336$ and $\nu_{cut} \approx 1.2$ THz for $N = 1002$. Both frequencies are marked as vertical lines in Fig. 2. $\nu_{cut}(N)$ coincides approximately with the frequency of the transverse acoustic excitation corresponding to the lowest q value which is determined by the size of the simulation box. To see that this is the case note that the lowest q values for $N = 336$ and $N = 1002$ are $q_{min} = 0.37$ Å^{-1} and $q_{min} = 0.26$ Å^{-1}, respectively. In comparison to that the q values

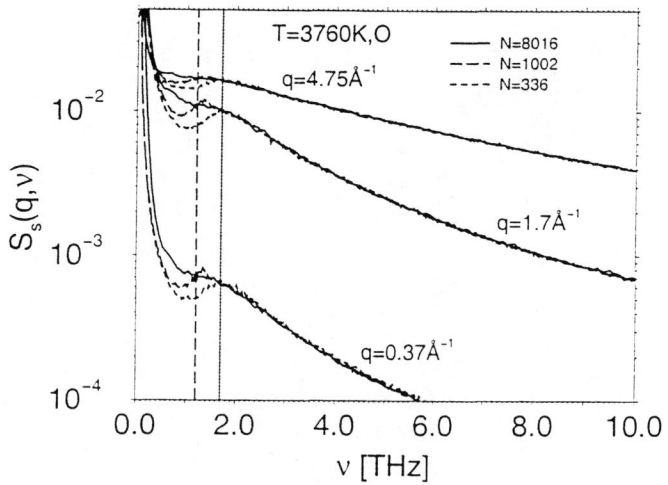

FIGURE 2. Self part of the dynamic structure factor as a function of frequency for the system sizes $N = 336, 1002$ and 8016 at the temperature $T = 3760$ K. For the explanation of the vertical lines see text.

we read off from the transverse acoustic dispersion branch for $T = 3760$ K and $N = 8016$, and which correspond to the frequency ν_{cut} for $N = 336$ and $N = 1002$, are 0.32 Å^{-1} and 0.22 Å^{-1}, respectively (see [7]). That the latter q values are slightly smaller than the corresponding values for q_{min} is due to the fact that the transverse dispersion branch has been determined from the peak maxima $\nu_{max}(q)$ in the transverse current correlation function. So there is always a significant contribution of transverse acoustic modes with frequencies $\nu < \nu_{max}(q)$ for a given q. All this can be summarized by saying that the absence of transverse acoustic modes is connected with a missing of excitations giving rise to the boson peak. Therefore, in the smaller systems only the high frequency part of the boson peak is present. Moreover, it seems that the boson peak modes are only fully present, for a given frequency, if there exist transverse acoustic excitations at the same frequency.

To learn more about the character of the boson peak excitations we varied also the masses of the silicon and oxygen atoms such that the total mass density remains fixed. Fig. 3 shows $S(q,\nu)$ at $T = 2750$ K and $q = 0.6$ Å^{-1} for the four mass pairs $M_1 = (28.086, 15.999)$, $M_2 = (14.043, 23.021)$, $M_3 = (44.085, 8.000)$, and $M_4 = (56.085, 2.000)$ where the first and the second number are the masses in atomic units for silicon and oxygen, respectively. Note that M_1 corresponds to the real masses of silicon and oxygen normally used in our simulation. From the figure we see that there is a strong dependence on the mass ratio for the two peaks visible for M_1 above 20 THz which are due to localized optical modes. In contrast to that, at least within the accuracy of the statistics of our data, there is no dependence for the modes giving rise to the boson peak and the acoustic modes which means

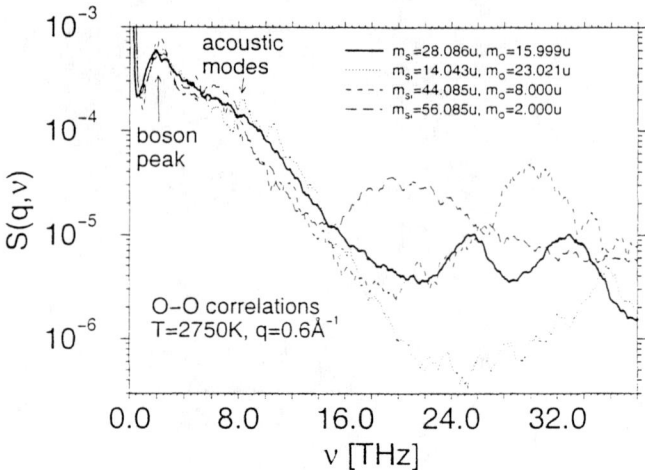

FIGURE 3. Dynamic structure factor at $T = 2750$ K and $q = 0.6$ Å^{-1} under variation of the masses of the silicon and oxygen atoms such that the mass density is fixed.

that the boson peak excitations cannot be strongly localized. This supports the aforementioned statement that the boson peak is due to the coupling to transverse acoustic modes. The fact that the boson peak is independent of the mass ratio between the silicon and oxygen mass is also a good test for theoretical models of the boson peak in silica.

In conclusion we investigated the excitations giving rise to the boson peak by means of molecular dynamics computer simulations. We find that the dynamic structure factor $S(q,\nu)$ scales roughly with temperature in the range 3760 K $\geq T \geq$ 300 K, which means that our silica system is in this sense quite harmonic even for temperatures as high as 3760 K. By calculating $S_s(q,\nu)$ for different system sizes we find that the modes contributing to the boson peak are only fully present at a given frequency if there exist transverse acoustic modes at the same frequency. This is supported by the fact that the height and the width of the boson peak are independent under the variation of the masses of silicon and oxygen if the mass density is fixed. So we observe that the boson peak is due to a coupling to transverse acoustic modes. Of course, the explanation of the nature of this coupling is an interesting goal for the future.

ACKNOWLEDGMENTS

This work was supported by BMBF Project 03 N 8008 C and by SFB 262/D1 of the Deutsche Forschungsgemeinschaft. We also thank the RUS in Stuttgart for a generous grant of computer time on the T3E.

REFERENCES

1. U. Buchenau, M. Prager, N. Nücker, A. J. Dianoux, N. Ahmad, and W. A. Phillips, Phys. Rev. B **34**, 5665 (1986); E. Rössler, A. P. Sokolov, A. Kisliuk, and D. Quitmann, Phys. Rev. B **49**, 14967 (1994); P. Benassi, M. Krisch, C. Masciovecchio, V. Mazzacurati, G. Monaco, G. Ruocco, F. Sette, and R. Verbeni, Phys. Rev. Lett. **77**, 3835 (1996); M. Foret, E. Courtens, R. Vacher, and J.-B. Suck, Phys. Rev. Lett. **77**, 3831 (1996).
2. Y. M. Gal'perin, V. G. Karpov, and V. N. Solov'ev, Sov. Phys. JETP **67**, 2386 (1988); W. Schirmacher, G. Diezemann, and C. Ganter, Phys. Rev. Lett. **81**, 136 (1998).
3. S. N. Taraskin and S. R. Elliott, Europhys. Lett. **39**, 37 (1997).
4. R. Dell'Anna, G. Ruocco, M. Sampoli, and G. Viliani, Phys. Rev. Lett. **80**, 1236 (1998).
5. J. Horbach, W. Kob, and K. Binder, J. Non-Cryst. Solids **235–237**, 320 (1998).
6. B. W. H. van Beest, G. J. Kramer, and R. A. van Santen, Phys. Rev. Lett. **64**, 1955 (1990).
7. J. Horbach, W. Kob, and K. Binder, to be published; J. Horbach, PhD Thesis, University of Mainz (1998).

Molecular Dynamics Simulation of Inelastic Neutron Scattering Spectra of Copper Azurin Hydration Water

Alessandro Paciaroni[a], Anna Rita Bizzarri[a,b] and Salvatore Cannistraro[a,b]

[a] Unitá INFM, Dipartimento di Fisica dell'Universitá,
Via Pascoli, I-06100 Perugia, Italy.
[b] Dipartimento Scienze Ambientali, Sezione Chimica e Fisica,
Universitá della Tuscia, I-01100 Viterbo, Italy.

Abstract. A molecular-dynamics (MD) simulation of hydrated Copper Azurin has been performed at different temperatures to study the single-particle dynamics of water surrounding the protein. The dynamical structure factor of hydration water shows a broad peak in the inelastic low-frequency region ($\sim 1.3\,\mathrm{meV}$). Such a peak, which is particularly evident at low temperatures, is reminiscent of the so-called boson peak, experimentally observed in several amorphous disordered materials. The reliability of the present MD simulations is further emphasized by the good agreement of the calculated dynamical susceptibility with that recently obtained by scattering experiments in similar systems.

INTRODUCTION

The structural organization and the temporal behaviour of both the protein and solvent at their interface are presently deserving a great interest in molecular biophysics; in fact, these aspects play a key role in understanding the interplay among structure, dynamics and functionality of proteins. It is well known that water, which is essential at least in a minimum amount in determining biological activity, influences protein mobility, folding and functionality [1]. In this connection, several recent studies have been devoted to elucidate the protein-water interface dynamical features, where a number of interesting and unusual phenomena occurs [2] such as the suppression of water crystallization in the first hydration shell of protein and its anisotropic and anomalous behaviour [2]. On the other hand, above a critical temperature, where protein motions show a departure from the harmonic temperature dependence, an increase of the hydration degree may activate dynamical transitions between nearly isoenergetic conformational substates, characterized by fast (β) and slow, collective, (α) relaxations [3,4], in analogy with what occurs in glassy materials. On such a ground, the nature of the mechanisms regulating the interplay between the spatio–temporal behaviour of both the solvent and protein, can be better clarified if the hydration water dynamical features are known. Molecular Dynamics (MD) simulation, which covers a time window from femto– up to several nano–seconds, could be a rewarding tool to provide an accurate microscopic information on the macromolecule-solvent system. Such an approach is particularly remarkable when used in conjunction with incoherent neutron scattering (INS) since both techniques probe the dynamical properties of a protein in the same time window. MD simulations can then be applied to estimate the INS intensities from the calculated atomic trajectories allowing one to make a direct comparison between experimental and simulated results. In the present study, MD simulation capabilities have been used to obtain rewarding information about the low-frequency behaviour of the hydration water of Azurin, a β–barrel copper protein involved in the electron transfer process. The results of such investigation are consistent with the presence, in the protein hydration water, of a typical spectral glassy anomaly that can be traced back to the so–called boson peak [5-7].

The MD trajectories of Az were generated by using the GROMOS96 program package [8] including the SPC/E potential for water [9]. Details on the simulations are reported in ref. [10]. The Az system at the required hydration level of h=0.40 g of H_2O/ g of protein (315 water molecules), was derived from the fully hydrated system. Cutoff radii of 8Å for the non-bonded interactions and of 14Å for the long-range charged interactions were used, respectively. MD trajectories of D_2O hydrated Az were followed for 600 ps at different temperature: T=100K, 180K, 220K, 300K. The stored trajectories of the hydration water hydrogen atoms were used to calculate the self intermediate scattering function $I(\mathbf{q},t)$ which can be derived through the relationship:

$$I(q,t) = 1/3N \langle \sum_{i=1}^{N} \exp[i\mathbf{q} \cdot (\mathbf{R}_i(t) - \mathbf{R}_i(0))] \rangle \tag{1}$$

where N is the number of water hydrogen atoms, $\mathbf{R}_i(t)$ is the position vector of the i-th atom at time t, and the brackets $\langle \rangle$ denote an averaging over both the water ensemble and the exchanged momenta $\mathbf{q}$ having the same modulus q, to take into account for anisotropic effects. The incoherent dynamical structure factor $S(q,\nu)$ can be directly calculated by the time Fourier Transform of $I(\mathbf{q},t)$.

RESULTS AND DISCUSSION

The low-frequency behaviour of the protein hydration water can be characterized by the incoherent dynamical structure factor $S(q,\nu)$. Such a quantity, calculated for the four temperatures at the fixed wave vector $q = 1.8\text{Å}^{-1}$, is shown in Fig. 1. A broad inelastic bump, centered around 1.3 meV ($\sim 11\,\text{cm}^{-1}$), is well visible up

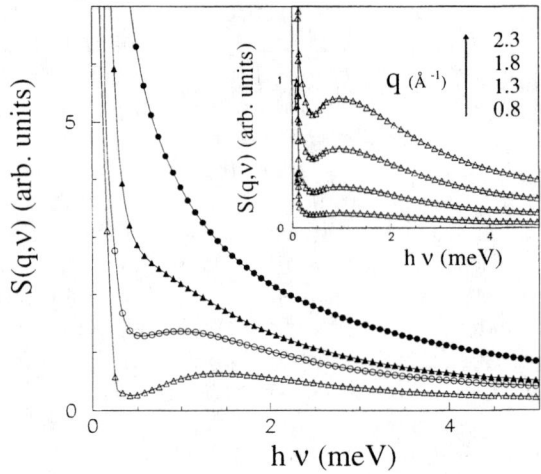

FIGURE 1. Incoherent dynamical structure factor at a fixed q value (q=1.8 Å^{-1}) for 100K (open triangles), 180K (open circles), 220K (closed triangles), and 300K (closed circles). Inset: $S(q,\nu)$ at T=100K for various wave vectors.

to 180K, but merges into the quasielastic contribution at 220K and 300K. Such an inelastic feature has been found, at approximately the same position, also in the Pc hydration water, with the same temperature dependence [11]. It should be remarked that the excess of inelastic scattering observed in the $S(q,\nu)$ of the protein hydration water, is reminiscent of the low-frequency inelastic feature revealed in the INS spectra of amorphous disordered materials, where it has been usually termed "boson peak" [5–7]. In glass-like systems, the inelastic peak has been observed in a range between 0.1 and 5 meV [5]. The origin of the boson peak observed in glasses, which is probably connected to certain low-temperature anomalies in their specific heat and thermal conduction [5], has been often attributed to some topological properties of the system [12]. It could be speculated that also in protein hydration water, as in glass-like systems [13,14], the boson peak is related to both structural correlations at intermediate range scale and topological disorder.

A direct comparison with INS experimental data could be a suitable test for the reliability of the MD results. To this aim the imaginary part of the dynamical susceptibility, $\chi(q,\nu) = \nu S_i(q,\nu)$, has been computed and has been compared with that recently measured in the hydration water of another protein (myoglobin [15]). The calculated χ is shown in Fig. 2 for the four different temperatures. At low temperatures, i.e. up to 220K, the curves are characterized by four peaks, each one corresponding to a well defined motion type of water molecules: the peak near 60

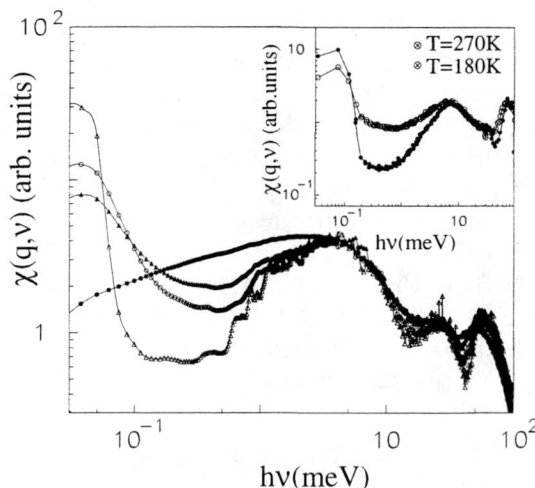

FIGURE 2. Az hydration water dynamical susceptibility, calculated in an extended frequency range ($q=1.8$ Å^{-1}). Symbols as in Fig. 1. <u>Inset</u>: experimental dynamical susceptibility of myoglobin hydration water (data from ref. [18].

meV is associated to librations, the intermediate peak at approximately 24 meV to fast translations, the broad one at 4 meV to flexing modes of three water oxygens, hydrogen-bonded to each other [16], and that at very low frequencies to slow (α) relaxational motions not resolved by the spectrometer. Hydration water dynamical susceptibility of Az shows a general trend which is in a good qualitative agreement with that obtained by neutron scattering at 270K and 180K for myoglobin hydration water [15] (see inset of Fig. 2).

The boson peak shown in $S(q,\nu)$ of Az hydration water may represent a fingerprint of the amorphous character of the system [5]. Moreover, the fact that the peak is no longer visible at 220K and 300K may suggest that hydration water has a "fragile" character (according to the Angell' s nomenclature [7]), as it has been supposed also for Pc [11]. Actually, in fragile liquids, the boson peak is present only below the glass–temperature T_g; strong glasses being characterized by the presence of the boson peak in all the temperature range. However, it should be observed that in our system the boson peak could be simply hidden by the increase of the quasielastic scattering contribution at 220K and 300K. Therefore the conclusion that protein hydration water has a "fragile" character requires further investigations and a detailed analysis of the quasielastic region to be really proved.

Interestingly enough, recent INS measurements performed on deuterium hydrated Az focussed on the inelastic behaviour of the protein itself, displayed also a bump centered at approximately 3meV and persisting up 220K [10,17]; also MD simulations on the same system provided an inelastic bump with the same temperature dependence, but peaking at about 1.5meV [10,17]. Recently, it has been suggested that the low frequency excess of modes in protein spectra could be linked to motions involving protein cross chain interactions, as being strongly modulated by the hydration water dynamics [18,19].

All these observations outline a scenario in which the protein–solvent coupling seems to play a crucial role in determining the dynamical features of the protein hydration water. It could therefore be speculated about a possible connection between the presence of the boson peak in the hydration water and the excess of low frequency anomalies in protein macromolecules; this aspect deserving a further experimental investigation which is in progress.

REFERENCES

1. *Protein-Solvent Interactions*, edited by Gregory R.B., (Marcel Dekker, New York, 1995).
2. Rocchi C., Bizzarri A.R., and Cannistraro S., *Phys. Rev. E* **57**, 3315–3323 (1998) and refs. therein.
3. Green J.L, Fan J., and Angell C.A., *J. Phys. Chem.* **98**, 13780–13790 (1994).
4. Doster W., Cusack S., and Petry W., *Nature* **337**, 754–756 (1989).
5. Frick B., and Richter D., *Science* **267**, 1939–1945 (1995).
6. Foley M., Wilson M., and Madden P.A., *Phil. Mag. B* **71**, 557–569 (1995).
7. Ediger M., Angell C.A., and Nagel S.R., *J. Phys. Chem.* **100**, 13200–13010 (1996).
8. Van Gunsteren W.F., and Berendsen H.J.C., *Groningen Molecular Simulation (GROMOS) Library Manual*, Biomos, Groningen, 1996.
9. Berendsen H.J.C., Grigera J.R., and Straatsma T.P., *J. Phys. Chem.* **91**, 6269–6271 (1987).
10. Paciaroni A., Stroppolo M.E., Arcangeli C., Bizzarri A.R., Desideri A., and Cannistraro S., *Eur. Biophys. J.* (1998) submitted.
11. Paciaroni A., Bizzarri A.R., and Cannistraro S., *Phys. Rev. E.* **57**, R6277–6280 (1998).
12. Elliott S.R., *Europhys. Lett.* **19**, 201–205 (1992).
13. Sokolov A.P., Kisliuk A., Soltwish M., and Quitmann D., *Phys. Rev. Lett.* **69**, 1540–1543 (1992).
14. Bermejo F.J., Criado A., and Martinez J.L., *Phys. Lett. A* **195**, 236–244 (1994).
15. Settles M., and Doster W., *Farad. Disc. Royal soc. of Chem.* **103**, 269–279 (1996).
16. Sceats M.G., and Rice S.A., *J. Chem. Phys.* **72** 3236–3247 (1980).
17. Paciaroni A., Bizzarri A.R., and Cannistraro S., *J. Mol. Liquids* (1998) submitted.
18. Diehl M., Doster W., Petry W., and Schober H., *Biophys. J.* **73**, 2726–2732 (1997).
19. Cusack S., and Doster W., *Biophys. J.* **58** 243–251 (1990).

Analysis of low-frequency motions in proteins by computer simulation and neutron scattering

Gerald R. Kneller and Konrad Hinsen

Centre de Biophysique Moléculaire (UPR 4301 CNRS)

Rue Charles Sadron

45071 Orléans Cedex 2

France

E-Mail: kneller@cnrs-orleans.fr / hinsen@cnrs-orleans.fr

ABSTRACT

The combination of molecular simulations and thermal neutron scattering is a powerful tool to analyze the dynamics of macromolecular systems on pico- to nanosecond time scales. The main goal is the assignment of parts of the neutron spectrum to specific molecular motions or to a certain type of motion. We show recent examples where Molecular Dynamics simulations and normal mode analysis are used to analyze quasielastic and inelastic neutron scattering spectra of myoglobin. Our analysis allows to obtain a general view of protein dynamics on the time scale of inelastic neutron scattering and extends earlier work on quasielastic scattering. We show also that normal mode analysis with much simplified force fields yields the vibrational density of states with about the same accuracy as full Molecular Dynamics simulations. This allows to obtain estimates of inelastic neutron spectra from simulation with low computational effort.

I INTRODUCTION

Thermal neutron scattering is a well-established technique to study the structure and dynamics of molecular systems at the atomic level. Neutron time-of-flight and backscattering spectrometers yield space- and time correlations of atomic positions on length scales between roughly 1 and 100 Å and time scales ranging from pico- to nanoseconds. These time and length domains are also covered by classical Molecular Dynamics (MD) simulations. In MD simulations one generates trajectories of typically several hundred or thousand particles by solving Newton's equations of motion numerically using an empirical force field. Usually these particles represent the atoms in the simulated system. Neutrons interact with the atomic nuclei in the sample via Fermi's zero-range pseudopotential, i.e. with the constituents of the simulated system [1]. Therefore a direct comparison between simulated and experimental neutron scattering spectra is possible. More precisely, the differential neutron scattering cross section can be expressed in terms of spatially Fourier transformed particle density correlation functions whose classical counterparts can be computed from the MD trajectories. In principle even quantitative comparisons are possible (see e.g. [4]) under the condition that almost classical systems are considered and recoil effects are not dominant [5]. In this way the empirical force field can be validated, and the simulated intensities allow at the same time a detailed analysis of the structure and dynamics of the system under consideration. Many examples for applications of MD simulations to the interpretation of neutron scattering spectra can be found in the literature. This technique is particularly important for complex systems for which an interpretation of the experimental spectra in terms of simple analytical models is difficult, if not impossible.

In this paper we show how the quasi-elastic and the inelastic part of neutron scattering spectra from globular proteins can be analyzed by using MD simulations and novel normal mode techniques, respectively. In section II we discuss the origin of diffusive motions in globular proteins as they are seen by quasi-elastic neutron scattering. The corresponding inelastic part of the neutron scattering spectra is discussed in section III. We use a normal mode calculation combined with projection techniques to analyze the types of the contributing motions and the corresponding spatial frequencies. In section IV we demonstrate that a much simplified protein force field, enabling rapid normal mode calculations, may be used to analyze protein vibrations over the whole frequency range. Concluding remarks are presented in section V.

II DIFFUSIVE MOTION

Neutron scattering experiments and crystallographic studies have revealed that protein dynamics and function are strongly dependent on temperature [2,3]. At very low temperatures the motion is described by harmonic vibrations around an equilibrium position which is determined by a local energy minimum. Many of these local minima and corresponding equilibrium positions ('substates') may exist. With increasing temperature the protein has enough kinetic energy to cross the energy barriers between different minima. The resulting diffusive motion can be studied by quasi-elastic neutron scattering. Such experiments for myoglobin have shown a relatively abrupt onset of diffusive motion at about 200 K [2]. Since little is known about the details of the energy landscape of proteins, one can only speculate about the type of the diffusional motion seen in the experiments. One can imagine jump diffusion between relatively well distinguished energy minima, or small-step diffusion of molecular subunits. MD simulations can be used to get more insight into the details of the motion. In [7] the following approach was used: We studied myoglobin, a globular protein with about 75% α-helical structure, and performed a standard MD simulation with a length of 300 ps without explicit solvent, representing the latter by a distance -dependent dielectric constant. For details we refer to [7]. In the analysis we looked at the diffusional motion of the protein side-chains. Since side-chains are the ends of the protein tree structure they have more freedom to perform stochastic motions at low energetic cost than the atoms in the protein backbone. To examine the side-chain motion, we extracted their global rigid-body motion from the simulated trajectory, creating a new trajectory in which only the rigid-body component of the side-chain motions was kept. Mathematically speaking, this component is obtained by finding for each time frame the optimal superposition of a given rigid reference structure with the corresponding structure in the MD configuration. This requires an efficient superposition algorithm which is described in [8] and implemented in the nMOLDYN analysis package [9]. We chose the side-chain configurations in the energy-minimized start structure as reference coordinates for the coordinate fits.

Figure 1 shows the comparison of the average atomic mean-square displacements of myoglobin computed from the full trajectory and the trajectory where only the rigid-body motion of the side-chains is considered. Both curves match well and exhibit to a large extent the characteristics of a 'liquid-like' small-step diffusion process according to the Einstein model. In this model, one obtains the well-known relation $< (\mathbf{R}(t) - \mathbf{R}(0))^2 > = 6Dt$ for the mean-square displacement, with $\mathbf{R}$ being the position of an atom and D the corresponding diffusion constant. Since the atomic motion in a protein is confined, the Einstein model can certainly not be valid on longer time scales. The fact that the mean-square displacement computed from the rigid-body trajectory exceeds the one computed from the full trajectory means that the hypothesis of rigid-body motion yields a slight overestimation of the diffusive side-chain motion. The inset in Figure 1 shows the simulated dynamic structure factors in comparison with experimental data. Since about 50% of the atoms in biological samples are hydrogen atoms and incoherent neutron scattering from hydrogen dominates by far all other scattering processes, Figure 1 shows effectively the incoherent dynamic structure factor of the hydrogen atoms. It should be mentioned that in the experimental spectra the vibrational part has been subtracted [2], but not in the simulation data. This explains the systematic deviation above 0.2 meV. In the quasielastic region below 0.2 meV, experimental and simulation data match perfectly, showing that the simulations describe well the molecular dynamics on the time scale of the experiment. The resolution of the experimental and simulated spectra is about 8 μeV and 25 μeV, respectively. The latter value is still much smaller than the width of the quasielastic line.

Summarizing the above study, we can say that diffusive motion in proteins on the picosecond time scale is due to liquid-like rigid-body motions of the protein side-chains. Although the side-chains also undergo internal conformational changes, the latter play only a minor role in the diffusion process. Our finding does not confirm the hypothesis of Doster *et al.* [2] who used a simple analytical model to explain the diffusion process in myoglobin. According to this model, the quasielastic scattering can be explained in terms of jump diffusion with a jump length of about 1.5 Å. Such jumps are seen in methyl rotations which are, however, hindered by a rotational barrier of about 3 kcal/mol and therefore occur rarely.

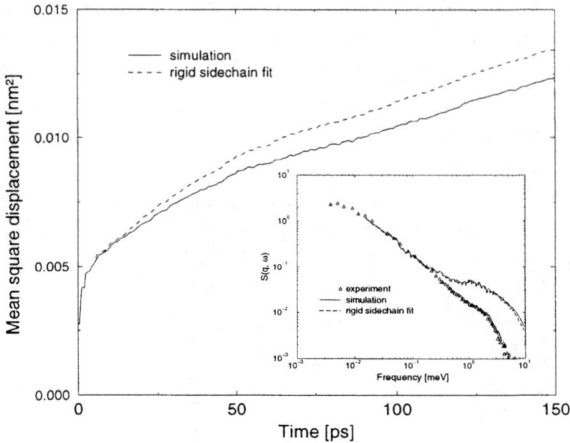

FIGURE 1. Mean square displacements of the hydrogen atoms in myoglobin. Solid line: Full trajectory. Broken line: Internal side-chain motion removed. **Inset:** Corresponding incoherent dynamic structure factors from simulation (solid and broken line) and experiment (triangles).

III VIBRATIONAL DYNAMICS

We now turn to the vibrational dynamics of proteins, which can be studied by *inelastic* neutron scattering. In this case the emphasis is not on the interpretation of experimental data, but on a description of the motions that are seen by inelastic neutron scattering. For this purpose we use standard normal mode analysis; in the next section, we will show that vibrational protein spectra can be obtained with reasonable accuracy even from a strongly simplified harmonic force field.

It is a well-known general feature of physical systems that fast motions are localized, involving only a few atoms that are close to each other, whereas slow motions describe large-scale deformations. This behavior is caused by the distance dependence of the relevant interactions: interactions between atoms at short distances, e.g. bonded atoms, are much stronger than the smooth long-range interactions between distant atoms. Mathematically it is expressed by a generally monotonic relation between spatial and temporal frequencies, known as a dispersion relation. A standard dispersion relation is not meaningful for proteins, because they are specific finite-size inhomogeneous objects. However, it is still of interest to study the relation between spatial and temporal frequencies for elastic vibrations of proteins.

Such an analysis can be performed by projecting each normal mode on a subspace that contains motions up to a specified wavenumber. Such a subspace has been introduced in Ref. [13] for the purpose of approximate normal mode calculations. We constructed these subspaces in wavenumber increments of 0.125π nm^{-1} and calculated the length of the projection of each normal mode onto each subspace. Since the length of a normal mode vector is one by definition (normalization), the square of the length of the projection can be interpreted as the fraction of the motion that is described by the subspace. Numerical differentiation with respect to wavenumber produces a two-dimensional spectrum as a function of wavenumber and frequency, which is shown in Figure 2.

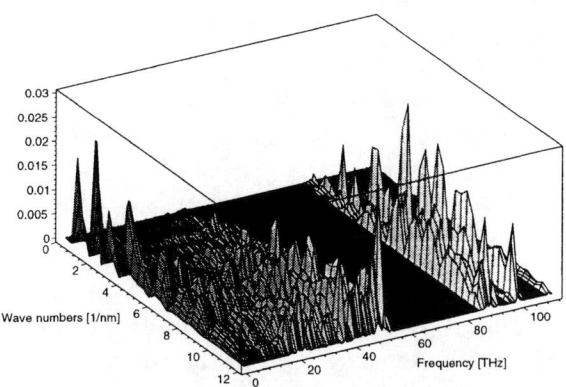

FIGURE 2. Spatial-temporal frequency spectrum of crambin. The discrete peak structure in the low-frequency low-wavenumber region is a consequence of the finite system size.

It is evident that up to $k \approx 3$ nm^{-1} only very low frequency motions are possible. The discrete structure of the spectrum in this region is a consequence of the finite size of the system; in between the peaks that are situated periodically along the k-axis there are no additional wavenumber vectors that could contribute. With increasing k, the low-frequency motions contribute less and the higher frequencies become more important. Of particular interest is the well-separated band describing the bond stretching motions of the hydrogen atoms. It starts to show important contributions at unexpectedly low k values (≈ 2 nm^{-1}), much lower than the onset of some other motions which have substantially lower frequencies. This shows that the hydrogen motions are not well described by a uniform collective motion. The picture is reminiscent of the normal mode structure of solids; the low-frequency band corresponding to the acoustic branch (for $k \to 0$ we find $\omega \to 0$ as well), and the high-frequency band corresponding to an optical branch (almost reaching $k = 0$ but remaining at a finite frequency).

To complement this study, we have made several further analyses, described in detail in Ref. [10], with the goal of understanding what kinds of motions occur in the low-frequency part of the spectrum, especially in the first peak located at ≈ 2 THz (60 cm^{-1}). We found that the largest contribution to this peak comes from motions in which the entire residues remain essentially rigid, i.e. deformations of secondary-structure elements and motions on an even larger scale, such as domain motions. The remaining part is caused mostly by rigid-body motions of the side chains relative to the backbone, i.e. the vibrational analog of the rigid-body side-chain motion that was found to be the dominant contribution to diffusive dynamics. Side-chain deformation occurs in a somewhat higher frequency range around 10 THz (300 cm^{-1}).

IV SIMPLIFIED FORCE FIELDS

An interesting question in the study of protein dynamics is how sensitive the dynamics is to details of the interactions. Comparisons of normal modes obtained from different force fields and different parameters for the long-ranged electrostatic interactions have shown that the low-frequency part of the spectrum is sensitive to changes in the treatment of electrostatics, e.g. cutoff methods [11]. Recent studies have demonstrated that the general shape of the spectrum describing the very beginning of the spectrum, as well as the motions that occur at these frequencies, are not sensitive to details of the force field at all and can be reproduced with simple harmonic interactions with a distance-dependent force constant [12,13]. At the other end of the frequency spectrum, the bond-stretching motions are essentially described by a single bond stretching term in the commonly used empirical force fields. It is therefore interesting to investigate which force field details are important for motions at intermediate frequencies.

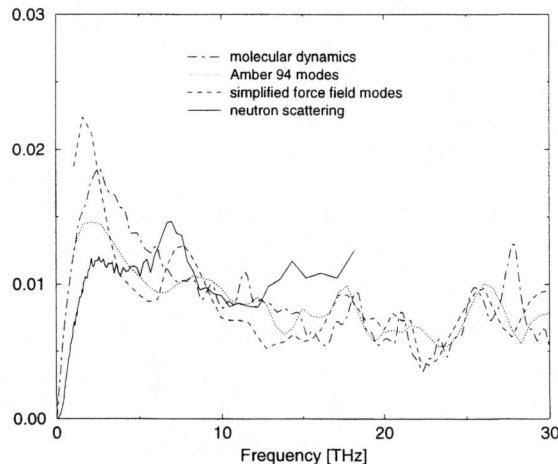

FIGURE 3. Frequency spectra from two normal mode calculations with different force fields compared to spectra from molecular dynamics and from neutron scattering.

We have found that the energy terms describing the covalent bond structure (bond stretching, bond angle and dihedral angle bending) are by themselves sufficient to describe the largest part of the spectrum, above ≈15 THz (450 cm^{-1}), and that bond stretching and bond angle bending are responsible for everything above ≈25 THz (750 cm^{-1}). Combining the three covalent bond structure terms with the simple harmonic force field that has proven sufficient for describing low-frequency domain motions [13] results in a very simple force fields which reproduces the whole frequency spectrum. Figure 3 shows the frequency spectrum from two normal mode calculations, one using the full Amber 94 force field [14] (without electrostatic cutoff) and one using the simplified force field just described, in comparison to a spectrum obtained from molecular dynamics [6] and an experimental neutron scattering spectrum [15]. The molecular dynamics spectrum was obtained using the CHARMM force field [16] with a cutoff of 0.8 nm smoothed by a switching function, and a united-atom model in which only polar hydrogens are represented explicitly.

The differences between the Amber 94 normal modes and the CHARMM dynamics trajectory show the influence of force field differences and anharmonic effects. It it clear that the deviations associated with the simplified force field stay within the range of deviations due to such effects. They mostly concern the low-frequency end of the spectrum, which is known to be sensitive to changes in the long-range interactions. Since none of the empirical force fields used in protein simulations, and particularly none of the popular cutoff schemes, was developed with the goal of an accurate description of low-frequency dynamics, there remains a large uncertainty as to what is the "correct" description of this frequency range. Our simplified force field thus cannot be said to be clearly worse than the detailed empirical force fields, and it has the advantage of simplicity and computational efficiency. In fact, a normal mode calculation using this force field does not require a lengthy energy minimization, and its second-derivative matrix is highly sparse due to absence of real long-range interactions. For most applications of normal mode calculations, we consider this force field sufficient and very useful.

The neutron scattering data shows more important differences from the various theoretical spectra. These are due several effects whose individual contributions are difficult to estimate: inaccuracies in the theoretical model, technical limitations in the experiments, and differences in the systems being studies (presence of solvent molecules etc.). Some of these differences could be eliminated by more elaborate calculations. Nevertheless, the comparison shows that the frequency scales and the general shape of the spectrum are the same, allowing an application of results obtained from interpretations of normal mode spectra to spectral data of different origin, such as the analysis discussed in the previous section.

V SUMMARY AND CONCLUSIONS

We have shown that Molecular Dynamics simulations and normal mode analyses are useful methods to reveal details in the dynamics of complex molecules that are not directly accessible by experimental techniques. They can nicely complement neutron scattering experiments in the quasielastic and inelastic regime. The first result is that small-step rigid side-chain diffusion dominates the quasielastic part of a protein spectrum. Although the example was myoglobin, we expect this result to be true for at least all globular proteins. The second result is that our simplified harmonic force field consisting of a contribution for large scale deformations and a contribution describing the covalent bond structure yields inelastic protein spectra with about the same accuracy as a full Molecular Dynamics simulation. This allows a quick computation of estimates for inelastic neutron spectra and shows that the dynamics around a stable equilibrium position does not require a detailed description of long-ranged interactions.

ACKNOWLEDGMENTS

We thank Dr. W. Doster for providing the neutron scattering data shown in Figure 3.

REFERENCES

1. S.W. Lovesey, 'Theory of Neutron Scattering from Condensed Matter', Vol. 1, Clarendon Press, Oxford, 1984.
2. Doster, W., Cusack, S. & Petry, W., *Nature* **337**, 754–756 (1989)
3. Rasmussen, B.F., Ringe, A.M. & Petsko, G.A., *Nature* **357**, 423–424 (1992)
4. Kneller, G.R. & Geiger, A., *Mol. Phys.* **70**, 465–483 (1990)
5. Kneller, G.R., *Mol. Phys.* **83**, 63–87 (1994)
6. Furois-Corbin, S., Smith, J.C. & Kneller, G.R., *Proteins* **16**, 141–154 (1993)
7. Kneller, G.R. & Smith, J.C., *J. Mol. Biol.* **242**, 181–195 (1994)
8. Kneller, G.R., *Mol. Sim.* **7**, 113 (1991)
9. Kneller, G.R., Keiner, V., Kneller, M., Schiller, M., *Comp. Phys. Comm.* **91**, 191–214 (1995) and Report ILL95KN02T, Institut Laue-Langevin, 156 X, F-38042 Grenoble Cedex, France
10. Hinsen, K. & Kneller, G.R., submitted
11. Teeter, M.M. & Case, D.A., *J. Phys. Chem.* **94**, 8091–8097 (1990)
12. Tirion, M.M., *Phys. Rev. Lett.* **77**, 1905–1908 (1996)
13. Hinsen, K., *Proteins* **33**, 417–429 (1998)
14. Cornell, W.D., Cieplak, P., Bayly, C.I., Gould, I.R., Merz Jr, K.M., Ferguson, D.M., Spellmeyer, D.C., Fox, T., Caldwell, J.W. & Kollman, P.A., *J. Am. Chem. Soc.* **117**, 5179–5197 (1995)
15. Settles, M. & Doster, W., In: Biological Macromolecular Dynamics, Eds. Cusack, S., Büttner, H., Ferrand, M., Langan, P. & Timmins, P., Adenine Press, New York, 1997
16. Brooks, B.R., Bruccoleri, R.E., Olafson, B.D., States, D.J., Swaminathan, S. & Karplus, M., *J. Comp. Chem.* **4**, 187–217 (1983)

Simulation of Inelastic Neutron Scattering Spectra for Water Ice

- A most effective way of testing water potentials.

Jichen Li* and John Tomkinson[+]

*Department of Physics, UMIST, PO Box 88, Manchester, M60 1QD, UK
[+] ISIS Facility, Rutherford Appleton Laboratory, Chilton, Didcot, Oxon, OX11 0QX, UK

Abstract: The vibrational spectra measured by inelastic neutron scattering techniques can be simulated by either lattice dynamic or molecular dynamic methods. Simulating the measured vibrational spectra of ices provides a test of the accuracy of the various water potentials. Our work demonstrates that the advantages and the disadvantages of each of simulation techniques. Combining their advantages, we were able to simulate a range of effects from these potentials.

Vibrational Spectra and Water Potentials

The quantitative studies of the properties of water and ice require detailed consideration of the forces acting on the atoms and the molecules. Experimental information about the strength of the H-bond interaction can be obtained directly by measuring vibrational spectra. A particular vibrational mode (or phonon) is determined by the interatomic force constants, which in turn are the double differentials of the potential function. Therefore, measuring dynamic properties constitutes one of the most powerful ways of investigating interatomic potentials in a given material. Such investigations are traditionally carried out by means of optical spectroscopy, such as IR absorption and Raman scattering. These are very powerful techniques, which have been highly refined, and their usage has resulted in extensive and valuable data for water and ice. In water and ice, however, the normal selection rules governing the interaction of radiation with matter are broken due to the local structural disorder (or proton disorder in crystalline structures of ice) and analysis of the spectra is difficult in general. On the other hand, although IR and Raman spectra are very sensitive to the intramolecular modes involving the O-H stretching and bending, they are less sensitive to the intermolecular modes involving the vibrations of water molecules against each other. Therefore, under normal circumstances, optical spectroscopy provides only limited data in the translational region, which is vitally important for obtaining direct information about the H-bond interaction.

The theoretical description of the hydrogen bonding in water and ice can be attributed to a potential function which is available in various forms, such as simple pair-wise SPC, MCY, TIP4P and polarisable SK potentials (see review articles 1,2). Choosing the appropriate experimental data for the validating tests of these potentials is therefore important. One of the conventional methods was to reproduce the partial radial correlation functions $G_{HH}(r)$, $G_{OO}(r)$ and $G_{OH}(r)$ for water obtained by neutron diffraction. In general, the molecular dynamic (MD) simulations of water structures using these potentials give good agreement with the ones obtained experimentally. It is often seen that the simple rigid point charge potentials, such as SPC, give almost identical results to the very complicated polarisable potentials as illustrated by Dang and Chang (3). In fact, the uncertainties (or errors) introduced in the partial correlation functions by the data reduction of the measured diffraction data were much greater than the errors (or differences) between the MD simulated data and the measured data. The systematic errors in the experimental data arise from two main sources: inelasticity and non-equivalence of H and D in the isotopic substitution (for details see ref. 1). In addition, the experimental measurements were made in reciprocal space (i.e. Q space) which requires Fourier transformation of the measured $S_{ij}(Q)$ to the real space variable $G_{ij}(r)$ in for comparison with the MD simulation results.

Since high resolution neutron spectroscopy became available in recent years, it has provided additional experimental data for benchmark testing. Both MD and lattice dynamic (LD) simulations provide the power spectrum (a sum of the normal modes) using the first and second derivative of the potential respectively. These simulations can be directly compared with the experimental spectrum without involving the Fourier transformation and other uncertainties (or errors) associated with diffraction data. Hence, simulating the inelastic neutron scattering (INS) spectra for a large variety of crystalline and amorphous phases of water ice is of considerable advantage in the process of examining possible potential functions.

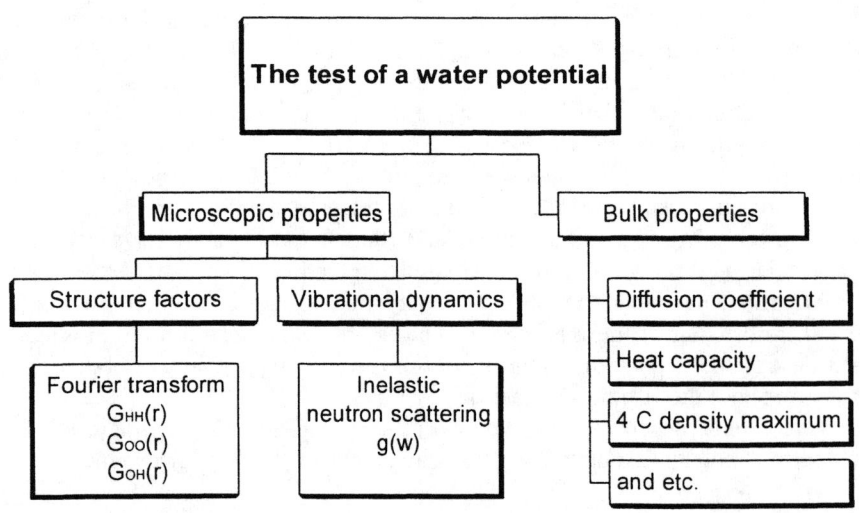

Lattice and Molecular Dynamic Simulations for Ice Ih

Using the classic water potentials, the INS spectrum for ice Ih was simulated using LD and MD techniques. The aims were two fold: first; to understand the fundamental reasons behind the doubling of the molecular optic peaks at 28 and 37 meV observed in the measured phonon density of states (i.e. the power spectrum) of normal ice Ih (see Fig. 1), and second; to determine which features, if any, could be produced by the existing literature potentials (1,2). The studies, therefore, provide a much-needed understanding of the vibrational dynamics arising from these potentials.

The simulation cell for the LD calculations used consists of 32 water molecules with hexagonal symmetry (ice Ih). The protons in the structure were disordered by the use of a random walk program (4). The structure is not quite large enough to fully represent the observed proton disordering, it however sufficiently represents the mixture of different configurations. Our earlier work, on a series of lattice sizes from 4-32 molecules, demonstrated that the result for 32 molecules (8 times of the primary unit cell) reproduces the measured spectrum very well (5).

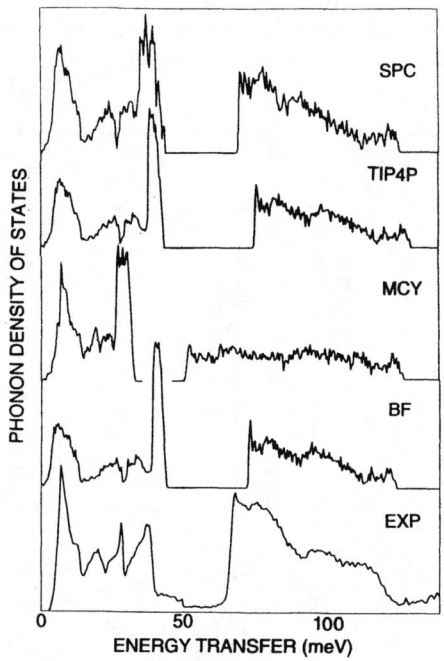

FIGURE 1. Calculated spectra for ice Ih using some classic pair-wise potentials. The bottom curve is the measured spectrum (at ~10 K) using TFXA at ISIS for comparison.

In this series of calculations, the lattice constants were initially set for $c = 7.32$ Å and $a = 4.50$ Å and the program relaxes both the molecular structure and lattice constants to minimise the total energy of the crystal for a given potential. The additional intramolecular strengthening (k = 36 eV/Å^2) and bending (g = 2.6 eV/Å^2) force constants were also introduced for all the rigid water potentials in order to stabilise the internal structure of the molecule, they also give reasonable intramolecular frequencies (4). As a consequence of the non-rigidity of the water molecule in these models, its dipole and quadruple moments are slightly larger than those given in other work. Here TIP4P provides a dipole value of 2.35 D, which is accidentally close to the experimental value of ~2.5 D, similar increases were also found for quadruple values. After relaxation, the dynamical matrix is resolved and integrals of the phonon modes across the first BZ were made. The plots of the phonon density of states (PDOS) for a few typical pair-wise potentials are shown in Fig. 1 and the curve at the bottom of the figure is the experimental data measured on TFXA at ISIS (6) for comparison.

In the translational region below 50 meV, the main features of the spectra for SPC, BF and TIP4P potentials are similar to spectra calculated using the simple force constant model (5). A single peak at less than 45 meV is predicted to dominate the molecular optic modes peaked at about 40 meV. For the MCY potential the molecular optic peak is shifted to considerably lower energy, 28.6 meV, indicating the hydrogen bonding of this potential is much softer than reality. The low energy cut-off for the librational band is also much lower than for the other potentials used. These general comments are confirmed by our MD simulations (see Fig. 2).

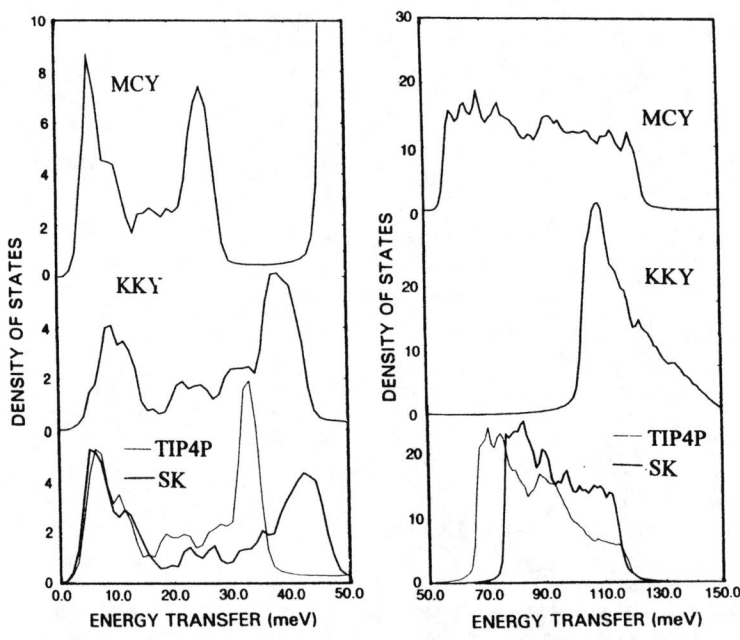

FIGURE 2. MD simulation results for ice Ih using MCY, KKY, TIP4P and SK potentials (7). The features shown in the figure are very similar to the LD results in Figure 1.

MD simulation of vibrational dynamics has considerable advantages over LD. This is because a variety of effects, such as long-range charge interactions, polarisation, anharmonicity and many-body terms are introduced naturally. Hence, it offers a viable alternative to calculate the INS spectrum for ice (i.e. the PDOS). For details of this technique and its application to PDOS calculations see ref (7,8).

Recently, there have been many MD simulations of the vibrational dynamics of ice using rigid, non-rigid or polarizable potentials (7). The resulting spectra (for a 512 molecular cell) show very similar features to LD results for the MCY and TIP4P potentials in the translational and librational regions. The spectrum calculated using the polarisable SK potential closely resembles the result from TIP4P in the energy transfer less than 20 meV as shown in Fig. 2. The principal difference is that the polarisation has increased the high energy cut-off of the translation from 36 meV for TIP4P to 47 meV for SK and broadened the peak considerably. This broadening phenomenon was our primary interest and was also observed from other polarisable potentials. This may imply that polarisation affects the strengths of H-bonds differently for different proton configurations (or relative molecular orientations), hence the orientational variations of the potentials are greater than the pair-wise ones, much as we would expect.

Although the aim of the simulations was to reproduce the INS spectrum for ice Ih the resulting spectra obtained from these potentials all fail to reproduce the INS spectrum. Specifically, the split optic modes of the translational region appear as a single feature in the calculated spectra. However, the simulations are model independent and both LD and MD results give broadly the same spectral features. This confirms the general reliability of the results presented. So far no potential, we have tested by either LD or MD, was capable of reproducing the measured spectrum for ice, and the differences remain fundamental. Simple potentials of the TIP4P type usually produce a single, narrow optic peak. However, there is clearly a tendency for polarisable potentials to broaden the optic features beyond that obtained from simple potentials. This broad peak indicates that the orientational variation of the potential function has been increased considerably but it may still be less than the values used in the force constant model (8,9). One would, therefore, expect that a better polarisable potential could eventually be able to reproduce the features in the INS spectrum.

REFERENCES

1. Franks, F, *Water Science Review*, Cambridge University Press, 1985, Vol 1.
2. G.W. Robinson, S.B. Zhu, S. Singh and M.W. Evans, *Water in Biology, Chemistry and Physics*, World Sciencitic, 1996.
3. Dang, L.X., and Chang, T-M., *J. Chem. Phys.* 106, 8149-8159 (1997).
4. Dong, S.L., *Ph.D. Thesis*, UMIST, Manchester, UK (1998).
5. Li, J.C., and Leslie, M., *J. Phys. Chem.* 101, 6237-6342 (1997) and 6304-6307 (1997).
6. Li, J.C., *J. Chem. Phys.* 105, 6733-6755 (1996).
7. Burnham, C.J., *Ph.D. Thesis*, Salford, UK (1998) & Burnham, C.J., Li, J.C., Leslie, M., and Xantheas, S.S., *J. Chem. Phys.* (1998), accepted.
8. Li, J.C., and Tomkinson, J., in *"Theoretical and Computational Chemistry"* eds. P.B. Balbuena and J.M. Seminario, Elsevier Science, 1998.
9. Li, J.C., and Ross, D.K., *Nature* 365, 327-329 (1993).

Extracting the vibrational density of states from neutron scattering data: beyond the effective density of states

S.R. Elliott, A. Haar, R.D. Oeffner and S.N. Taraskin

Department of Chemistry, University of Cambridge, Lensfield Rd., Cambridge CB2 1EW, UK

Abstract. The connection between the true vibrational density of states and that measured in inelastic neutron experiments is derived for multicomponent systems. The resulting correction function is related to the relative partial vibrational densities of states. For two structural models (vitreous silica and vitreous germania), this function is calculated and used for a comparison of the calculated and the experimental vibrational densities of states.

Inelastic neutron scattering is widely used for investigating the atomic dynamics of condensed systems. One of the important vibrational characteristics of solids available from inelastic neutron scattering experiments is the vibrational density of states, $g(\omega)$. It can be extracted from an analysis of the dynamical structure factor, $S(\mathbf{Q},\omega)$ [1,2]. For a single component system, in the incoherent approximation, the vibrational density of states is directly proportional to the dynamic structure factor averaged over a certain $\mathbf{Q}$-range. This is not the case for a multicomponent system where the vibrational density of states and the averaged dynamical structure factor are connected by a correction function, $C(\omega)$, which depends on frequency.

The correction function depends on atomic masses, neutron scattering lengths, Debye-Waller factors, and relative partial vibrational densities of states. Also, this function depends on the $\mathbf{Q}$-range used for averaging. In order to calculate $C(\omega)$, all these characteristics need to be known, wherein the main difficulty lies in the calculation of the relative partial vibrational densities of states and the Debye-Waller factors. One of the possible ways to obtain these characteristics is to create numerically a structural model of the system, e.g. by means of standard or *ab initio* molecular dynamics. Then a normal-mode analysis can be done to calculate the necessary vibrational characteristics.

We demonstrate how this can be done by considering two glassy systems: vitreous silica (v-SiO_2) and vitreous germania (v-GeO_2). The calculated correction function is found to vary appreciably, so that its deviation, $\delta C(\omega) = C(\omega) - C_s$ ($C_s \equiv 1$ is the value for a single-component system), can be of the same order as C_s, $|\delta C(\omega)| \sim 1$

and this has to be taken into account when comparing the numerically calculated, that is the true, vibrational density of states and the experimental vibrational density of states [2,3].

The dynamic structure factor measured experimentally is proportional to the generalized VDOS, $g_{\text{gen}}(\mathbf{Q},\omega)$, so that the correction function connecting the true and generalized VDOS is $C(\mathbf{Q},\omega) = g_{\text{gen}}(\mathbf{Q},\omega)/g(\omega)$. It separates into a coherent and an incoherent part, $C(\mathbf{Q},\omega) = C_{\text{incoh}}(\mathbf{Q},\omega) + C_{\text{coh}}(\mathbf{Q},\omega)$. The coherent part oscillates strongly with Q and effectively is reduced to zero when averaged over Q, so that in the incoherent approximation for isotropic systems the correction function becomes the following (see Ref. [2] for more details):

$$C(Q,\omega) \simeq 1 + \frac{A(Q)}{3} \sum_\alpha \frac{\overline{b_\alpha^2}}{m_\alpha} \exp\{-2\overline{W_\alpha}(Q)\} \left(\rho_\alpha(\omega) - \rho_\alpha^{(0)}\right) \equiv 1 + \delta C(Q,\omega_j) \quad (1)$$

Here $A(Q) = 3\sum_i m_i / \sum_i \overline{b_i^2} \exp\{-2\overline{W}_i(Q)\}$ ($\overline{b}_i$ and m_i are the neutron scattering length and mass of atom i, respectively), $\exp\{-2\overline{W_\alpha}(Q)\} = \exp\{-Q^2\langle u_\alpha^2\rangle/3\}$ is the Debye-Waller factor ($\langle u_\alpha^2\rangle$ is the mean squared displacement of the atoms of type α), and $\rho_\alpha(\omega)$ is the relative partial vibrational density of states, which can be expressed in terms of the eigenvectors, $\mathbf{e}_i(\omega)$, of the dynamical matrix as $\rho_\alpha(\omega) = \sum_{i\in\alpha} |\mathbf{e}_i(\omega)|^2$. In the limiting case of low temperatures and small values of Q, as the Debye-Waller factor is close to unity, the correction function $C(Q,\omega)$ becomes Q-independent. In the opposite limiting case of high temperatures and large values of Q, the correction function, $C(\omega) \simeq (\sum_i m_i / \sum_{i\in\alpha} m_i) \rho_\alpha(\omega)$ (α stands for the species characterized by the smallest atomic displacements) is also Q-independent. In other cases, after averaging over Q, $C(Q,\omega)$ becomes Q-independent.

The structural models of v-SiO$_2$ and v-GeO$_2$ were created by standard molecular dynamics (NPT ensemble) using interatomic pair potentials based on *ab-initio* calculations for small atomic clusters [4–6]. The melts were quenched from $T = 5000$K (v-SiO$_2$) and $T = 3000$K (v-GeO$_2$) to the relaxed glassy state at $T \sim 10^{-4}$K with an average quench rate of 1K/ps. The dynamical matrices in the relaxed state were directly diagonalized, resulting in all eigenvectors and eigenfrequencies. A detailed analysis of v-SiO$_2$ has been presented in Ref. [2], so that the main attention below is paid to v-GeO$_2$.

We have calculated the correction function for the constructed structural models of v-SiO$_2$ and v-GeO$_2$ and found that they have indeed the same shape as the relative partial VDOS, as expected in the incoherent approximation (see Fig. 1).

The correction function depends on the interval of momentum transfer, $Q_{\text{min}}^{\text{av}} \lesssim Q \lesssim Q_{\text{max}}^{\text{av}}$, over which the averaging of the dynamical structure factor is performed. In the case of v-SiO$_2$, e.g. for $Q_{\text{min}}^{\text{av}} \sim 1 - 6\text{Å}^{-1}$ and $Q_{\text{max}}^{\text{av}} \sim 12 - 18\text{Å}^{-1}$, the correction function is close to that correction function, $C_0(\omega)$, in which the Debye-Waller factor is replaced by unity (see Fig. 1c) and the dependence on $\{Q_{\text{min}}^{\text{av}}, Q_{\text{max}}^{\text{av}}\}$ is not significant. For v-GeO$_2$, the correction function depends appreciably on the interval $\{Q_{\text{min}}^{\text{av}}, Q_{\text{max}}^{\text{av}}\}$, being close to unity after a shift of this interval to higher values of Q (see Fig. 1d). This behaviour is related to the larger values of the mean squared atomic displacements in v-GeO$_2$ as compared to v-SiO$_2$ which are due to the differences in the relative partial vibrational densities of states in these materials (cf. Figs. 1a and 1b), so that the Debye-Waller factor cannot be replaced by unity in v-GeO$_2$.

We have used the correction function, presented in Fig. 1d, to compare the simulated VDOS for the structural model of v-GeO$_2$ (the dashed line in Fig. 2) with our experimental data (the circles in Fig. 2) obtained from an inelastic neutron scattering experiment performed on MARI ($T \simeq 20K$, $Q^{av}_{min} \simeq 8 Å^{-1}$ and $Q^{av}_{max} \simeq 16 Å^{-1}$, 2 – 35THz) and on IN6 ($T \simeq 300K$, $Q^{av}_{min} \simeq 0.5 Å^{-1}$ and $Q^{av}_{max} \simeq 2.6 Å^{-1}$, 0.25 – 2THz). The effective VDOS, $g_{eff}(\omega) = C(\omega)g(\omega)$, which can be compared with the experimental data, is shown by the solid line in Fig. 2. As can be seen, the main features of the effective VDOS (range, peak positions) are reasonably well reproduced, which gives additional support for the quality of the interatomic pair potential used in the simulations (see also Ref. [6]). The remaining discrepancies we ascribe to sabtle deficiencies of the potential that we have not unmasked yet.

We have shown that, in order to compare the effective vibrational density of states measured in inelastic neutron scattering experiments with the true one found, e.g. from simulations, a correction function should be used, which depends on the wavevector averaging range and is approximately proportional to the relative partial

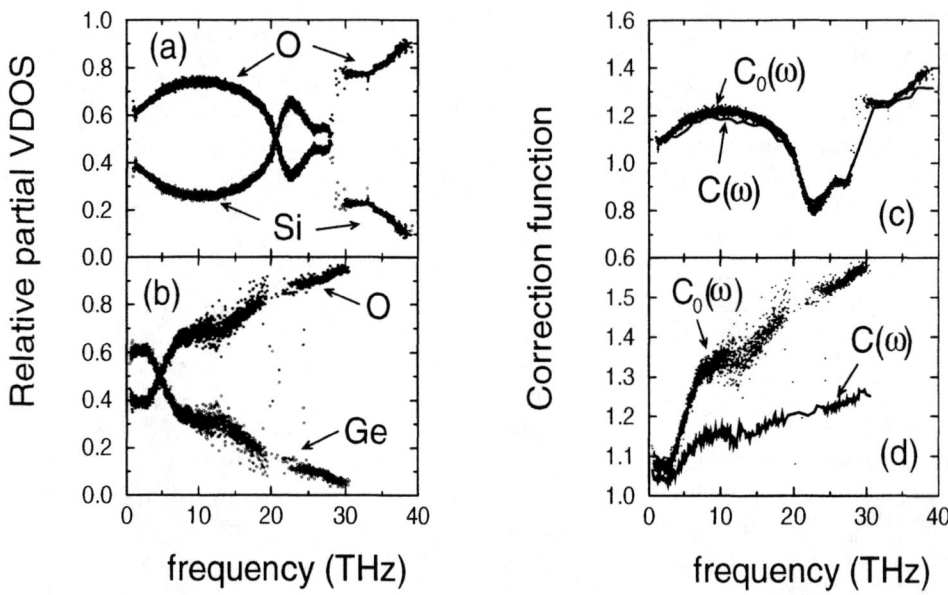

FIGURE 1. The relative partial VDOS for Si and Ge atoms and O atoms in (a) v-SiO$_2$ and (b) v-GeO$_2$. The correction function $C_0(\omega)$, i.e. the one with the Debye-Waller factor set to unity and the correction function $C(\omega)$ for (c) v-SiO$_2$ for $Q^{av}_{min} = 0.22 Å^{-1}$ and $Q^{av}_{max} = 12 Å^{-1}$ and (d) v-GeO$_2$ for $Q^{av}_{min} = 6 Å^{-1}$ and $Q^{av}_{max} = 12 Å^{-1}$.

vibrational density of states. This function has been calculated for two glassy systems, and we found that the correction function can depart from unity by up to 20% for v-SiO$_2$ and 60% for v-GeO$_2$. Therefore, if the simulated VDOS is found, e.g. from a normal-mode analysis or from the velocity-velocity auto-correlation function in MD, before comparing it with the generalized VDOS available from an inelastic neutron scattering experiment, it should be multiplied by the correction function which can also be calculated numerically. On the other hand, if the generalized VDOS is found experimentally it should be divided by the correction function. In order to do this, a numerical structural model has to be at hand including interatomic potentials, relaxed structure and normal modes.

REFERENCES

1. D.L.Price, and K.Sköld, in *Neutron Scattering*, Part A, Eds., K.Sköld, and D.L.Price, (Academic press, London 1986).
2. S.N. Taraskin and S.R. Elliott, Phys.Rev., B **55**, 117, (1997).
3. J. Dawidowski, F.J. Bermejo, and J.R. Granada, Phys.Rev., B **58**, 706, (1998).
4. S.Tsuneyuki, M.Tsukada, H.Aoki, and Y.Matsui, Phys.Rev.Lett., **61**, 869 (1988).
5. B.W.H.Van Beest, G.J.Kramer, and R.A.Van Santen, Phys.Rev.Lett., **64**, 1955 (1990).
6. R.D. Oeffner and S.R. Elliott, Phys.Rev., B **58**, 14791 (1998).

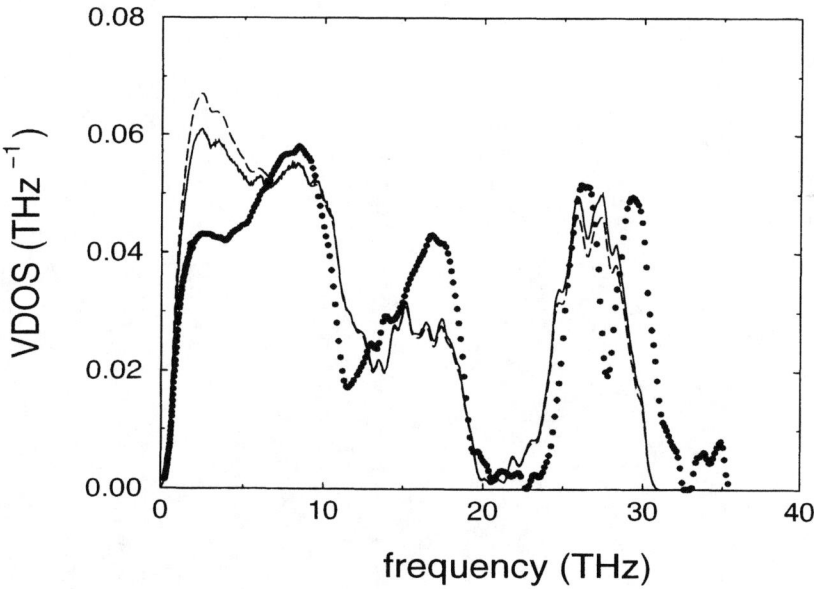

FIGURE 2. The VDOS for v-GeO$_2$: experimental (circles), calculated (dashed line) and calculated and weighted with the correction function (solid line).

INTERNAL VIBRATIONS AND TUNNELING

EXAFS Calculations using Debye-Waller factors deduced from inelastic neutron scattering

Matthew L. Hanham and Robert F. Pettifer

Department of Physics, University of Warwick, Coventry CV4 7AL, United Kingdom

Abstract. An EXAFS calculation has been performed for the zinc tetraimidazole molecular cluster, which includes thermal damping corrections established from inelastic neutron scattering. Analytic expressions for the damping factors involving multiple scattering path variances are discussed and some variances shown. Explicit calculation of thermal parameters enables a quantitive comparison between experiment and theory to be carried out, without the need for fitting. For one particular code (FEFF6), differences between theory and experiment are evident.

INTRODUCTION

EXAFS (extended X-ray absorption fine structure) is a much used technique, popular with physicists, chemists, and biologists for examining the local structure around a metal centre in unknown structures, such as glasses, liquids, catalysts and metallo-enzymes.

Structural information can be obtained from examination of EXAFS data including bond lengths, coordination numbers and thermal vibrational parameters. There are two main techniques used to analyse experimental EXAFS spectra which can be described as comparative with a standard and fitting an *ab initio* theoretical model.

Comparative techniques do not always work owing either to the non availability of an appropriate standard, or the non separability of a standard spectrum into its component scattering signals from which unknown spectra can be synthesised. If comparative techniques are appropriate then they are the technique of choice. Unfortunately, recourse has frequently to be made to *ab initio* theory to solve the bulk of problems.

Ab initio theory at present, relies on a variety of approximations, which have not been tested. A major reason for this is that full vibrational information is not usually available for many materials and consequently has to be treated with adjustable parameters. This masks the effects of the approximations in the *ab initio* theory. It is in this context that we have worked with inelastic neutron scattering

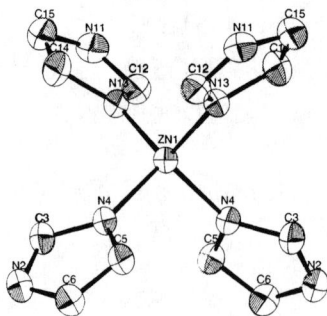

FIGURE 1. The structure of the cation of crystalline zinc (II) tetraimidazole tetrafluroborate at 150K. Note that the thermal ellipsoids here are relative to the lattice points in the crystal and are of the same magnitude for all atoms. In comparison, the relative zinc-nitrogen bond length variance is an order of magnitude smaller than the variance of the atom position with respect to the lattice points.

and used CLIMAX [1] to establish the vibrational eigenvalues and vectors of the molecular cluster zinc tetraimidazole (Fig 1) [2,3]. We have used this data to test EXAFS theory for one main code MSXAS [4].

Here we present further consideration of the calculation of the damping factors resulting from thermal vibrational disorder. An EXAFS calculation, including the vibrational data, is performed using the code FEFF6 [5], for the molecular cluster.

THERMAL DAMPING THEORY

We make the initial assumption that the thermal motion of the constituent atoms of our model compound follow, to a good approximation, harmonic oscillations and choose mass weighted normal coordinates (Q) as a set of generalized coordinates. Following Benfatto et al. [6] we write the pth contribution to the EXAFS signal for a static lattice as

$$\chi_p(k) = A_p(k, R_p) \sin(kR_p + \phi(k, R_p)), \qquad (1)$$

where p runs over all paths including MS (multiple scattering). $A_p(k, R_p)$ contains the number of atoms and a scattering form factor. $\phi(k, R_p)$ is a path dependent phase factor which involves the emitter and scatterer phase-shifts. The temperature distribution of the coordinates is given by

$$P(r)\,dr = \frac{e^{-\frac{(r^\dagger M^{-1} r)}{2}}}{(2\pi)^{\frac{N}{2}} \det[(M)^{\frac{1}{2}}]}\,dr, \qquad (2)$$

in which M is the correlation function given by

$$M_{ii} = \left(\frac{\hbar}{2\omega_i}\right) \coth\left(\frac{\hbar\omega_i}{2K_bT}\right), M_{ij} = 0 \; \forall \; i \neq j. \tag{3}$$

The quantum statistical average of the pth MS contribution (Eqtn 1) is given by expanding A and $\psi = kR_p + \phi(k,r)$ as Taylor series about equilibrium positions [6], giving

$$\langle \chi_p(k) \rangle = A_0 \left(1 + \frac{(A_1^\dagger M \psi_1)^2}{A_0^2}\right)^{\frac{1}{2}} e^{-\frac{(\psi_1^\dagger M \psi_1)}{2}} \sin\left(kR_p + \phi_0 + \frac{(A_1^\dagger M \psi_1)}{A_0}\right), \tag{4}$$

where A_1, ϕ_1 and below, R_{p1}, are derivatives with respect to the normal coordinates and $\dagger$ denotes transpose. Of the three additional terms in Eqtn 4, compared with Eqtn 1 the largest difference is contributed by $e^{-\frac{(\psi_1^\dagger M \psi_1)}{2}}$ in which the term $e^{-\frac{(R_{p1}^\dagger M R_{p1})}{2}}$ involving the path is most important. The expression $R_{p1}^\dagger M R_{p1}$ is the variance of the path length and in the case when the path has only two legs (single scattering) the expression $e^{-\frac{(R_{p1}^\dagger M R_{p1})}{2}}$ can be identified with the more familiar expression for damping involving a Debye-Waller like term $e^{-\frac{(R_{p1}^\dagger M R_{p1})}{2}} = e^{-2\sigma_p^2 k^2}$, where σ_p^2 is the variance of the bond length enclosed in the path p.

CALCULATION OF THERMAL DAMPING

The starting point for the calculation of the thermal damping terms is the normal mode analysis of the model compound performed on data from inelastic neutron scattering investigations. The normal mode analysis was carried out as described in [3] using the software CLIMAX. The eigenvalues and relative eigenvectors (normal modes in terms of mass weighted cartesian displacements) were extracted from CLIMAX. The correlation function was calculated according to Eqtn 3. The path length was expressed in terms of mass weighted normal coordinates using the transformation from mass weighted cartesian displacements to mass weighted normal coordinates obtained from CLIMAX:

$$R_p = \sum_{i=1}^{n} \left[\sum_{j=1}^{3}\left[\sum_{k=1}^{3N-6}\left(\frac{l_{(i+1)jk}}{\sqrt{m_{(i+1)}}} - \frac{l_{ijk}}{\sqrt{m_i}}\right)Q_k - (a_{(i+1)_j} - a_{i_j})\right]^2\right]^{\frac{1}{2}} \tag{5}$$

where the path is represented by the list of n atoms indexed by i, the 3 cartesian components indexed by j, m_i is the mass of atom i, the transformation from mass weighted cartesian displacements denoted by l_{ijk}, a_{i_j} is the jth cartesian component of the equilibrium position of atom i and k denotes the normal mode. The derivative of the path length with respect to the normal coordinates was evaluated at $Q = 0$ giving

$$R_{p_1,s}|_{Q=0} = \sum_{i=1}^{n} \left[\sum_{j=1}^{3}(a_{i_j} - a_{(i+1)_j})^2\right]^{-\frac{1}{2}} \times \sum_{j=1}^{3}(a_{i_j} - a_{(i+1)_j})\left(\frac{l_{(i+1)_j s}}{\sqrt{m_{(i+1)}}} - \frac{l_{i_j s}}{\sqrt{m_i}}\right) \quad (6)$$

where $R_{p_{1,s}}$ is the sth element of the $3N - 6$ element vector R_{p_1}.

PATH LENGTH VARIANCES AND THE EXAFS SIGNAL

TABLE 1. Calculated half path-length variances, σ_T^2, at two temperatures for five of the most important multiple-scattering paths for photoelectrons emitted from the zinc atom in the cluster.

MS path	Half path-length variance, σ_T^2 (10^{-3}Å^2)	
	$T = 20K$	$T = 300K$
$Zn-N_{13}-Zn$	2.63	4.29
$Zn-N_4-Zn$	2.63	4.31
$Zn-N_{13}-C_{15}-Zn$	3.37	7.15
$Zn-C_{12}-N_{11}-N_{13}-Zn$	3.03	5.54
$Zn-C_3-N_2-N_4-Zn$	3.07	5.68

The EXAFS signal was calculated by FEFF6. FEFF6 generates individual scattering signals which are combined to give the total EXAFS signal. The path length variances were calculated for the paths chosen by FEFF6 as being most significant and fed to FEFF6 as Debye-Waller factors, weighting the contributions from each path to the total signal. 461 paths were considered and the variances calculated of which 316 were included by FEFF6 in the final calculation. Some path length variances are shown in Table 1, calculated at $20K$ and $300K$, the paths are defined according to the tabulation in Fig 1. A comparision between the calculated damped spectrum and an experimentally derived spectrum is shown in Fig 2.

DISCUSSIONS AND CONCLUSIONS

From Fig 2 it is clear that there is rough agreement between the experimental and calculated EXAFS, which improves with energy. There are also considerable differences. At the low energy end (3 to 10 Å^{-1}) there is a phase difference that is initially almost 1 Å^{-1}.

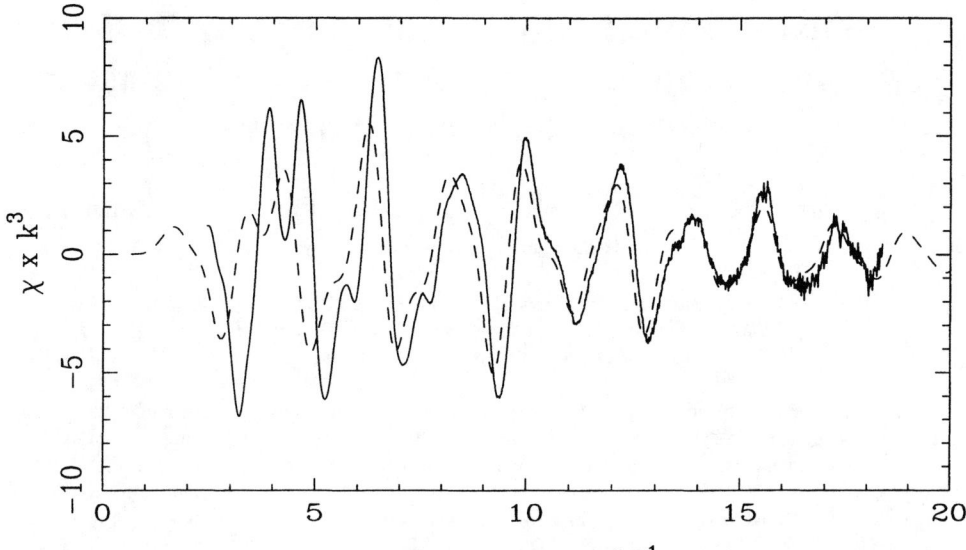

FIGURE 2. A comparison of the experimental (solid line) EXAFS and a calculated (dashed line) EXAFS including treatment of thermal disorder, of zinc tetraimidazole at 20K. There is good agreement at the high energy range (13-30 Å^{-1}) of amplitude and phase with the experimental EXAFS. At lower energies there is considerable overdamping and a phase shift of almost 1 Å^{-1}.

Concentrating on the high energy part of the data, the agreement between calculated and measured path length variance is better than ±4%. However, now we have a parameter free calculation and inadequacies in the scattering potentials can no longer be masked by changing thermal factors. We can now iterate on our potentials which can then be assumed to be transferable to unknown structures.

REFERENCES

1. Kearley G. J., *J. Chem. Soc. Faraday Trans. 2* **82**, 41–48 (1986).
2. Loeffen P.W., Pettifer R.F., Fillaux F., and Kearley G.J., *J. Chem. Phys.* **103**, 8444–8455 (1995).
3. Loeffen P.W., Pettifer R.F., and Tomkinson J., *Chemical Physics* **208**, 101–118 (1996).
4. Loeffen P.W. and Pettifer R.F., *Phys. Rev. Lett.* **76**, 636–639 (1996).
5. Rehr J.J., Zabinsky S.I., and Albers R.C., *Phys. Rev. Lett.* **69**, 3397–3400 (1992).
6. Benfatto M., Natoli C.R., and Filipponi A., *Phys. Rev. B.* **40(14)**, 9626–9635 (1989).

Density Functional Theory and *Ab Initio* Methods Applied to the Analysis of Inelastic Neutron Scattering Spectra

A. Navarro[*], M. Fernández-Gómez[*], J.J. López-González[*], F. Partal[*], J. Tomkinson[H] and G. Kearley[≅]

[*]*Department of Physical and Analytical Chemistry, University of Jaén, 23071 Jaén, Spain*
[H]*ISIS Rutherford Appleton Laboratory, Chilton, OX11 0QX, U.K.*
[≅] *Institut Laue Langevin, BP 156X, 38042 Grenoble, France*

Abstract. The INS spectra of several azines have been analyzed starting from force constant matrix calculated via *ab initio* and DFT methods. In order to reproduce both vibrational frequencies and intensities, the initial force constants must be fitted until the spectral profile is reproduced. We have checked the various ways which can be followed by the user in this kind of analysis. The choice between scaling or refinement-procedure of the force constants can make an important difference in the reproduction of the INS spectrum.

INTRODUCTION

At present we are witnessing an explosive growth in the ability of calculations to reproduce the observed dynamical behaviour of molecular systems. In fact, most experimentals papers now corroborate their results with such calculations. One of the commonest applications of computational chemistry is prediction of the vibrational spectrum, not only vibrational frequencies, but also the band intensities. Due to the greater use of optical spectroscopies, i.e. Infrared and Raman, many improvements have been directed towards the reproduction of these spectra.

However, the experimental measurements and the theoretical calculation of IR/Raman intensities are not yet considered to be fully reliable. INS represents an important advance in this matter because the spectral intensity can be predicted with considerable accuracy. At this point we introduce the *interaction* between INS and theoretical chemistry because INS intensity can be calculated from the cross section of each atom and the atomic displacements in the vibration, the latter being calculated via semiempirical, *ab initio* or most recently, Density Functional Theory (DFT) (1-2).

Although many DFT methods are *ab initio*, for historical reasons we will reserve the latter term for methods based on the wavefunction approach to the energy eigenvalue of the Schrödinger equation, i.e. HF and MP methods, as opposed to density functional methods that use electronic density to obtain the energy eigenvalue.

METHODOLOGY

One of the first steps in INS analysis is to choose of the appropriate level of theory with which to calculate the initial force-constant matrix. In this sense, computational chemistry offers a many options for the description of the Hamiltonian operator and molecular wavefunction. Over the last five years, the reproducibility of INS spectra has been tested from *ab initio* calculations, at HF and MP2 levels using various basis sets (3,4). More recently, the same procedure has been followed starting from DFT calculations (5). As expected, the results show that the introduction of electron correlation in the Hamiltonian, such as the MP2 and DFT methods, improves the quality of calculated vibrational frequencies as compared with those obtained with the HF method. In addition, the computatonal economy of the DFT methods make them a promising alternative approach to the study of many chemical systems which are too large for conventional correlated *ab initio* methods (1, 2).

Although the vibrational wavenumbers calculated via MP2 or DFT are close to those observed, the match between observed and calculated INS intensities is often far from perfect. Therefore, the next step is an optimisation of the initial force-constant matrix until the difference between observed and calculated spectra is minimized. There are two methods by which this can be achieved, both being valid. Firstly, the initial force-constant matrix is scaled following the standard procedure by Fogarasi and Pulay (6). Alternatively, the force constants are refined directly to the observed INS spectral profile by a mean-squares procedure (7). As we will demonstrate later, the refinement procedure has proved to be more convenient in reproducing wavenumbers as well as spectral intensities.

Finally, and independently of the option chosen, various coordinate-systems can be used to solve the secular equation, starting from the simplest valence internal coordinates, up to the independent symmetry coordinates, or even a combination of both, for instance, the 'natural coordinates' by Pulay et al. (6). In all cases, the procedure is the same, but obviously, the symmetrization of the secular equation significantly reduces the number of variables, and is therefore the method of choice.

In this paper, we will analyze the possibilities outlined above, showing the advantages and disadvantages of each option. The program used for this kind of analysis was CLIMAX (8), which provides a versatile tool for analysis of the INS spectra. The *ab initio* and DFT calculations have been made using the Gaussian/94 package program (9).

RESULTS

In this section we will summarize the results of systematic studies with two azines. The first is for the pyrimidine molecule. In Figure 1 we show its INS spectrum calculated directly from the atomic displacements matrix at the level MP2/6-31G*, and compare this with the experimental spectrum. Figure 2 shows the analogous spectrum for the pyridine molecule at the level B3LYP/6-31G*, i.e. using Becke's 3-parameter exchange functional (10) in combination with the Lee-Yang-Parr correlation functional (11) as implemented in Gaussian 94, with the 6-31G* basis set.

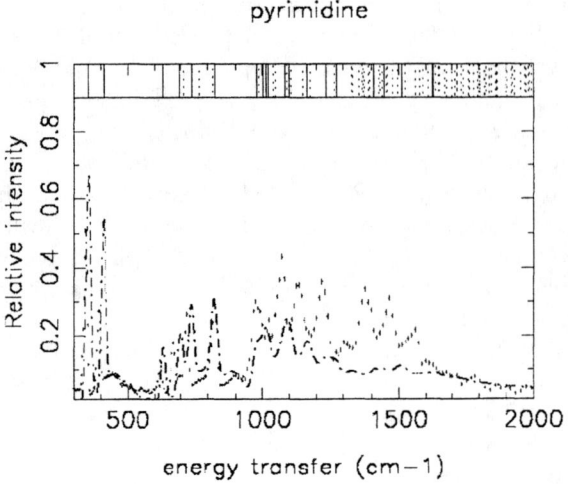

FIGURE 1. Observed (bars) and calculated (dot-dash) spectra for the pyrimidine molecule at the MP2/6-31G* level.

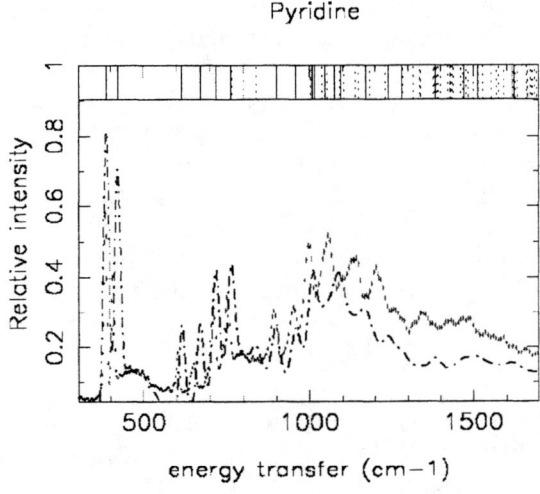

FIGURE 2. Observed (bars) and calculated (dot-dash) INS spectra for the pyridine molecule at the B3LYP/6-31G* level.

In both cases, the agreement between observed and calculated wavenumbers (maxima of the bands) is good, but the calculated intensities are clearly in error. Thus, we must fit some force constants until they not only reproduce the correct frequencies, but also the correct vibrational amplitudes. In all our studies, we use a constrained refinement procedure based on frequencies and intensities rather than a scaling procedure based on frequency-data alone. The advantages of this approach can be clearly illustrated with the following example. In Figure 3 we show the INS spectrum for the pyridine molecule in the region of B_1 out-of-plane normal modes after a first refinement of the B3LYP/6-31G* initial force constants. In this step, Set 1, all diagonal force constants, and a single off-diagonal, $F_{2,4}$, were refined. Their values are collected in Table 1, with the corresponding wavenumber information in Table 2. The refinement leads to a clear improvement of the vibrational frequencies, but the intensity for the v_3 mode is not correctly reproduced (see Figure 3).

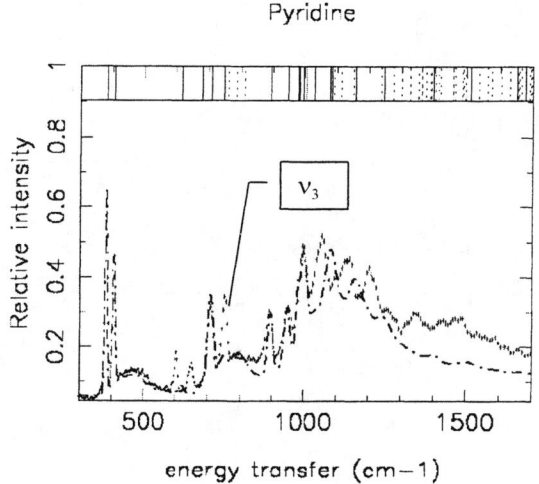

FIGURE 3. Observed (bars) and calculated (dot-dash) INS spectra from the refined force constants (Set 1) for the B_1 symmetry block.

To overcome this discrepancy, we refined up to three different off-diagonal force constants, i.e. $F_{1,4}$ $F_{2,4}$ and $F_{3,4}$, and as can be seen in Figure 4, the INS spectrum is correctly reproduced for all vibrations this symmetry block (see Figure 4). The final force constants (Set 2) and the corresponding wavenumbers are shown in Table 1 and 2, respectively.

TABLE 1. Symmetry force constants for the pyridine molecule (mdyn/Å)

Diagonal	$F_{1,1}$	$F_{2,2}$	$F_{3,3}$	$F_{4,4}$	$F_{5,5}$
B3LYP/ 6-31G*	0.304	0.287	0.337	0.334	0.302
Set 1	0.303	0.283	0.380	0.278	0.303
Set 2	0.275	0.285	0.449	0.276	0.287

Off-diagonal	$F_{1,2}$	$F_{1,3}$	$F_{1,4}$	$F_{1,5}$	$F_{2,3}$	$F_{2,4}$	$F_{2,5}$	$F_{3,4}$	$F_{3,5}$	$F_{4,5}$
B3LYP/6-31G*	-.011	0.003	0.025	-.009	-.029	0.025	-.002	-.028	-.032	0.062
Set 1	-.011	0.003	0.025	-.009	-.029	-.032	-.002	-.028	-.032	0.062
Set 2	-.011	0.003	-.009	-.009	-.029	-.051	-.002	-.036	-.032	0.062

Inspection of Table 1 reveals that in order to reproduce the wavenumbers and the spectral intensities, two off-diagonal force constants, $F_{1,4}$ and $F_{2,4}$, have reversed their signs from their starting values. Clearly, a scaling procedure could not bring about this change of sign, without which the INS spectra cannot be reproduced.

TABLE 2. Observed and calculated wavenumbers for the pyridine molecule (cm^{-1})

	Observed	B3LYP/6-31G*	Set 1	Set 2
v_1	1007	1017	1007	1007
v_2	948	959	948	948
v_3	748	765	748	748
v_4	710	720	710	710
v_5	406	421	406	406

We emphasise that we have used 'independent symmetry coordinates' which drastically reduces the number of independent parameters to refine. In the simplest analysis, when the whole secular equation is symmetry blocked, the number of possible interactions among the diagonal force constants because only those in the same symmetry block are possible. In Figure 5 we show the final INS spectrum refined from the MP2/6-31G* for the pyrimidine molecule.

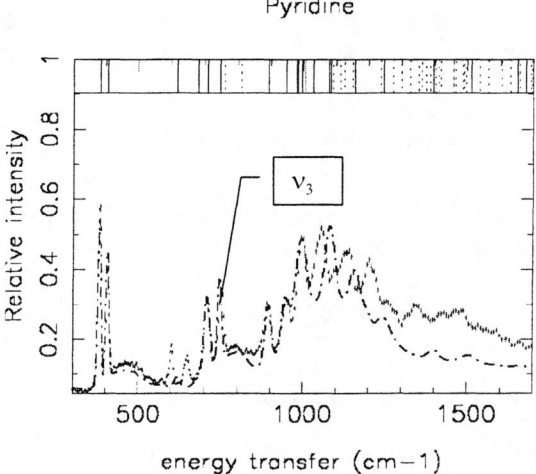

FIGURE 4. Observed (bars) and calculated (dot-dash) INS spectra from the refined force constants (Set 2) for the B_1 symmetry block.

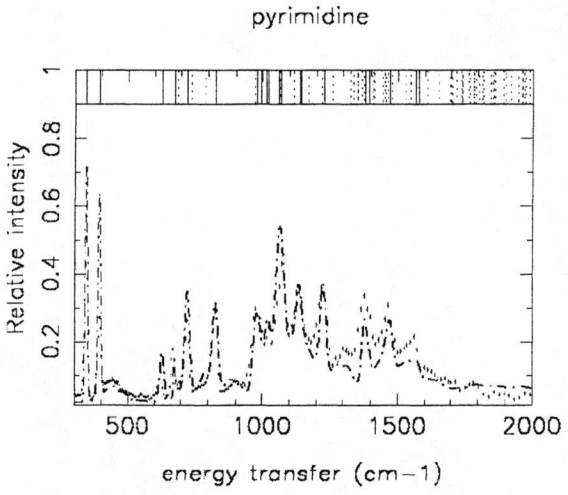

FIGURE 5. Observed and refined INS spectra for the pyrimidine molecule from the MP2/6-31G* calculation.

In summary, MP2 and DFT calculations seem to be a good starting point in INS analysis. In addition, if the symmetrization of the secular equation is performed, we can obtain an accurate reproduction of the INS spectrum refining a small number of parameters in the force constants matrix.

ACKNOWLEDGMENTS

We would like to thank Dr. Emilio Martínez, University of Castilla-La Mancha, Spain, for his help with some of the computer programs used in this work.

REFERENCES

1. Levine, I.N., *Quantum Chemistry*, New Jersey: Prentice Hall, 1991.
2. Seminario, J.M., Politzer, P., *Moderns Density Functional Theory. A tool for Chemistry*, Amsterdam: Elsevier, 1995.
3. Navarro, A., López González, J.J., Kearley, G., Tomkinson, J., Parker S.F., and Sivia, D.S, *Chem. Phys.*, **200**, 395 (1995).
4. Kearley, G., Tomkinson, J., Navarro, A., López González J.J., and Fernández Gómez, M., *Chem. Phys.*, **216,** 323 (1997).
5. Navarro, A., Fernández Liencres, M.P., Fernández Gómez, M., López González, J.J., Tomkinson, J., and Kearley, G. (to be published).
6. Fogarasi, G., and Pulay, P., *Vibrational Spectra and Structure*, Durig, J.R. (ed.), vol. 14, Amsterdam: Elsevier, , 1985.
7. Califano, S., *Vibrational State*, London: Wiley Interscience, 1976.
8. Kearley, G., *J. Chem. Faraday Trans. II*, **82**, 41 (1986).
9. Gaussian 94, Inc., Pittsburg PA, 1994.
10. A.D. Becke, *J. Chem. Phys.*, **98**, 5648 (1993).
11. C. Lee, W. Yang, R.G. Parr, *Phys. Rev. B*, **41**, 785 (1988).

Search for a Reliable Nucleic Acid Force Field Using Neutron Inelastic Scattering and Quantum Mechanical Calculations: Bases, Nucleosides and Nucleotides

Nicolas Leulliot[1], Hervé Jobic[2], Mahmoud Ghomi[1*]

[1]*Laboratoire de Physicochimie Biomoléculaire et Cellulaire, UPRESA 7033, Université P. & M. Curie, Case Courrier 138, 75252 Paris Cedex 05, France*
[2]*Institut de Recherche sur la Catalyse, 2 avenue A. Einstein, 69626 Villeurbanne, France*

Abstract. Neutron inelastic scattering (NIS), IR and Raman spectra of the RNA constituents: bases, nucleosides and nucleotides have been analyzed. The complementary aspects of these different experimental techniques makes them especially powerful for assigning the vibrational modes of the molecules of interest. Geometry optimization and harmonic force field calculations of these molecules have been undertaken by quantum mechanical calculations at several theoretical levels: Hartree-Fock (HF), Moller-plesset second-order perturbation (MP2) and Density Functional Theory (DFT). In all cases, it has been shown that HF calculations lead to insufficient results for assigning accurately the intramolecular vibrational modes. In the case of the nucleic bases, these discrepancies could be satisfactorily removed by introducing the correlation effects at MP2 level. However, the application of the MP2 procedure to the large size molecules such as nucleosides and nucleotides is absolutely impossible, taking into account the prohibitive computational time needed. On the basis of our results, the calculations at DFT levels using B3LYP exchange and correlation functional appear to be a cost-effective alternative in obtaining a reliable force field for the whole set of nucleic acid constituents.

INTRODUCTION

Ribo-nucleic acid (RNA) is an important biomolecule with a wide range of functions: transfer of the genetic code and genomic regulation, protein synthesis as well as catalytic properties (1). The study of the relation between structure and function in RNA is a challenging task. From the structural point of view, several physical techniques such as NMR, X-ray diffraction and vibrational spectroscopy permit the elucidation of conformational properties of RNA in solid and/or aqueous phases. RNA is a linear polymer of repeating units called *nucleotides* which are mainly composed of one of the four major nucleic acid bases (uracil, cytosine, adenine or guanine) attached to a five membered ribose ring (sugar) and a phosphate group. The flexibility of the RNA chain allows it to fold back onto itself, giving rise to very complex tertiary structures containing double helical and loop regions. Recent investigations (2) have shown that the nucleosides (base+sugar) can adopt different conformations:

C3'endo/anti, C2'endo/anti and C3'endo/syn. C2'endo and C3'endo refer to two different types of sugar puckering, while anti and syn designate the orientation of the base relative to the sugar moiety.

In order to comprehend the RNA structural properties by means of vibrational spectroscopy, Raman and IR spectroscopy in aqueous phase have been used in order to analyze the conformation of nucleotides by means of specific vibrational marker bands, whose wavenumber and/or intensity are conformation dependent (3). To fully understand and interpret these vibrational markers, we have also undertaken a systematic analysis of the vibrational spectra of bases, mononucleosides and mononucleotides by different experimental and theoretical techniques: NIS, Raman and IR spectroscopy and quantum mechanical calculations. We describe here a part of our recent results on the mononucleosides.

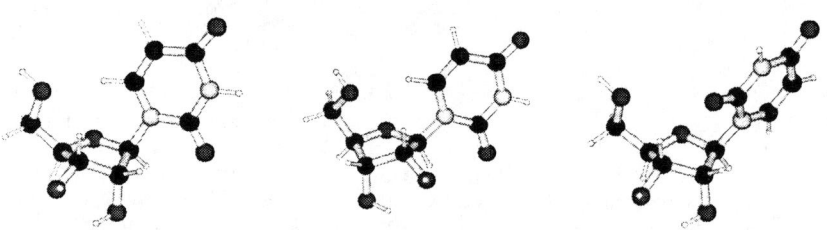

FIGURE 1. Optimized geometries of the three major conformers of uridine. From left to right, C3'endo/anti, C2'endo/anti, C3'endo/syn conformations. Note that the first two conformers correspond to a degenerate energy level while the last one presents an energy increase of +2.446 kcal/mol with respect to the other two.

MATERIAL AND METHODS

NIS spectra of the powder samples of the four bases as well as uridine (uracil+sugar) have been recorded at 20 K on TFXA spectrometer at ISIS neutron spallation source at the Rutheford-Appleton Laboratory (UK). All other NIS spectra (of cytidine, guanosine and adenosine) have been analyzed at 20 K at the Institut Laue-Langevin on IN1-BEF spectrometer. Theoretical calculations at HF, MP2 (only for the bases) as well as at the DFT/B3LYP levels were performed with the Gaussian94 code on Cray C-90 platforms. Geometry optimization as well as harmonic force field calculations were carried out using double-zeta polarized split valence basis sets with non-standard exponents for d-orbital polarisation functions (6-31G$^{(*)}$). The home-made program BORNS made possible the obtention of harmonic force field in terms of internal coordinates, redundancy removal, as well as vibrational wavenumber and atomic displacement calculations.

RESULTS AND DISCUSSION

Geometry optimization as well as vibrational calculations on the nucleic acid bases have been performed at HF+MP2 level of theory and published recently (4). In the case of the large size molecules such as mononucleosides and mononucleotides, we resorted to DFT calculations. The description of the calculated results concerning all the nucleosides exceeds the scope of the present paper, we limit here our discussion on uridine. Figure 1 shows the optimized geometries for the three conformers of uridine. On the basis of these results the C3'endo/anti and C2'endo/anti conformations have approximately the same electronic energy ($\Delta E=0.002$ kcal/mol), whereas the C3'endo/syn conformation presents a $\Delta E=+2.446$ kcal/mol raise in energy. Thus a change in the sugar puckering costs a negligible amount of energy while the anti to syn rotation of the base is accompanied by a notable energy increase. Consequently, contrarily to common belief, there is no preference at the nucleoside level for the C3'endo/anti conformation. Results from geometry optimization on uridine-monophosphate seem to indicate that the presence of the phosphate group leads the nucleotide to preferentially adopt a C3'endo/anti conformation.

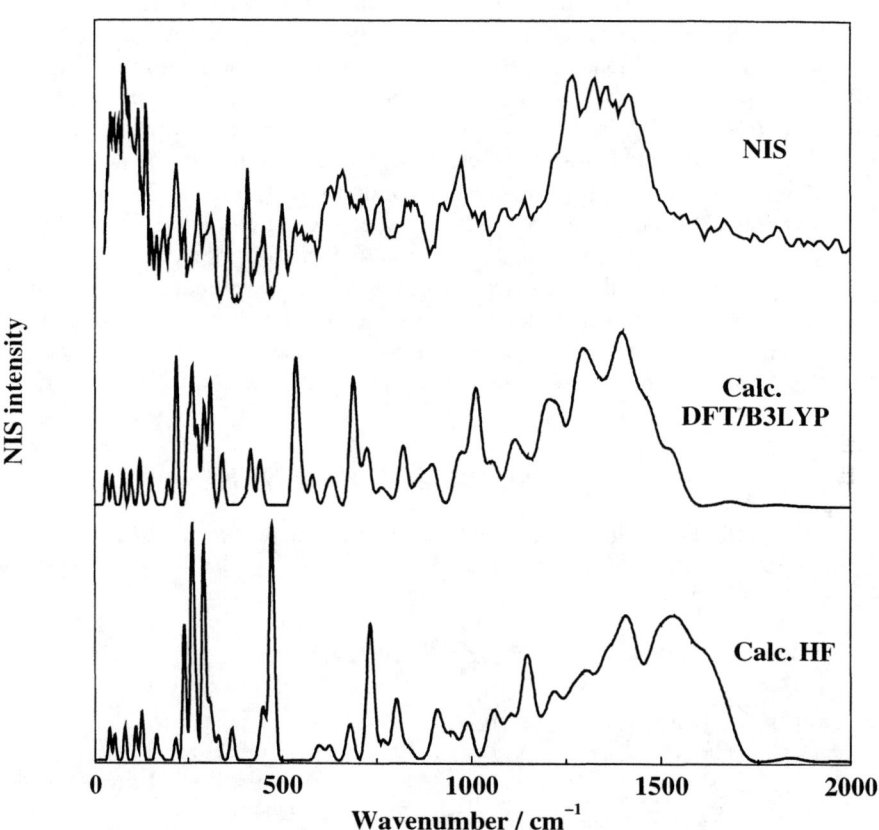

FIGURE 2. Comparison between the experimental and calculated (with the unscaled force fields obtained at HF and DFT levels) NIS spectra of uridine. Note that the lattice modes give rise to additional bands below 120 cm^{-1} which are not reproduced by calculations based only on internal modes.

Harmonic force filed calculations at HF and DFT levels showed that the energy order mentioned above is conserved after addition of the zero point vibrational energies. Uridine in crystal phase adopts a C3'endo/anti conformation (5). Thus, in Figure 2, we have compared the NIS spectrum obtained from powder samples with the first-order calculated spectrum of the C3'endo/anti conformer. It can be concluded that (i) the vibrational modes calculations at HF level do not reproduce the spectral data obtained from NIS (Fig. 2), IR and Raman (not shown here) spectroscopy. Even the application of large scaling factors (not considered in this work) seems to be insufficient to improve the agreement between the theoretical and experimental data. (ii) DFT calculations show very good overall agreement for both wavenumbers and intensities even without employing any scaling procedure (Fig. 2).

In conclusion, the advantage of NIS spectroscopy is evident in the analysis of nucleic acid constituents because of the affinity of neutrons to probe proton dynamics in inelastic process. We have shown in our investigations that this technique can be employed as a complementary tool to traditional optical vibrational spectroscopy (Raman and IR). In fact in a first step, NIS spectra allowed us to obtain valuable information on the vibrational modes of the bases in which the hydrogen displacements are included, especially the vibrational modes arising from C-H and N-H wagging motions (800-950 cm^{-1}). In a second step, the comparison between the NIS spectra from the bases and nucleosides allowed the vibrational modes of the sugar to be detected. One can mention here the intense bands located in the 1270-1400 cm^{-1}, 900-1000 cm^{-1}, 600-700 cm^{-1} and 250-350 cm^{-1} spectral regions (Fig. 2) which all arise from the angle-bending and torsional motions of the sugar pucker.

Our aim is also to use the harmonic force constants estimated for the four major nucleotides, in classical approaches such as molecular mechanics, molecular dynamics and Monte Carlo-Metropolis method, allowing the theoretical search of the conformational space of single- or double-stranded oligonucleotides, to be carried out.

ACKNOWLEDGEMENTS

The authors would like to thank the technical staff of the RAL (UK) and ILL (France) for NIS spectroscopy. We acknowledge the IDRIS (CNRS) for computational facilities on Cray C-94 and C-98 supercomputers.

REFERENCES

1. Cech, T. R., *Science* **236**, 1532-1538 (1987).
2. Varani, G., *Annu. Rev. Biophys. Biomol. Struct.* **24**, 379-404 (1995).
3. Abdelkafi, M., Leulliot, N., Baumruk, V., Bednarova, L., Turpin, P.Y., Namane, A., Gouyette, C., Huynh-Dinh, T., Ghomi, M., *Biochemistry* **37**, 7878-7884 (1998).
4. Aaamouche, A., Ghomi, M., Grajcar, L., Baron, M.H., Romain, F., Baumruk, V., Stepanek, J., Coulombeau, C., Jobic, H., Berthier, G., *J. Phys. Chem. A* **101**, 10063-10074 (1997).
5. Green, E.A., Rosenstein, R.B., Shiono, R., Abraham, D.J., Trus, B.L., Marsh, R.E., *Acta Cryst. B* **31**, 102-107 (1975).

NIS, IR and Raman Spectra with Quantum Mechanical Calculations for Analyzing the Force Field of Hypericin Model Compounds

Jozef Ulicny[1,2], Nicolas Leulliot[1], Lydie Grajcar[3], Marie-Hélène Baron[3], Hervé Jobic[4] and Mahmoud Ghomi[1*]

[1]*Laboratoire de Physicochimie Biomoléculaire et Cellulaire, UPRESA 7033, Université P. & M. Curie, Case Courrier 138, 75252 Paris Cedex 05, France*
[2]*Department of Biophysics, Safarik University, Jesenna 5, 04154 Kosice, Slovakia*
[3]*Laboratoire de Dynamique, Interactions et Réactivité, CNRS, 2 rue H. Dunant, 94320 Thiais, France*
[4]*Institut de Recherche sur la Catalyse, 2 avenue A. Einstein, 69626 Villeurbanne, France*

Abstract. Geometry optimization as well as harmonic force field calculations at HF and DFT levels of theory have been performed in order to elucidate the ground state properties of anthrone and emodin, two polycyclic conjugated molecules considered as hypericin model compounds. NIS, IR and FT-Raman spectra of these compounds have been recorded to validate the calculated results (geometry and vibrational modes). Calculated NIS spectra using the lowest energy conformers are in agreement with experiment. In addition, the intramolecular H-bonds in emodin predicted by the calculations can be evidenced using IR spectra as a function of temperature.

INTRODUCTION

Hypericin (Fig. 1) is a polycyclic aromatic dione which can be found in plants of the *Hypericum* genus. It can be considered as the main member of antiviral and antiretroviral agents used against several viruses including HIV [1-2]. It is known that the irradiation of hypericin with visible light leads to the production of singlet oxygen or free radicals [3-4]. On the other hand, this irradiation produces an important local pH drop. In addition, hypericin and its analogues can interact with several cellular targets like membranes, proteins and nucleic acids [5].

The direct analysis of the ground state properties (structure and vibrations) of hypericin is a difficult task, because this molecule is composed of eight conjugated rings with methyl and hydroxyl substitutions. This fact led us to undertake a step by step analysis of some other polycyclic compounds which are considered as model compounds of hypericin, i.e. anthrone and emodin (Fig. 1). Only the structural data of anthrone has been determined by X-ray diffraction in solid phase [6]. The molecular structure of emodin is deduced from the anthrone one by methyl and hydroxyl substitutions, together with the replacement of a methylene by a carbonyl bond.

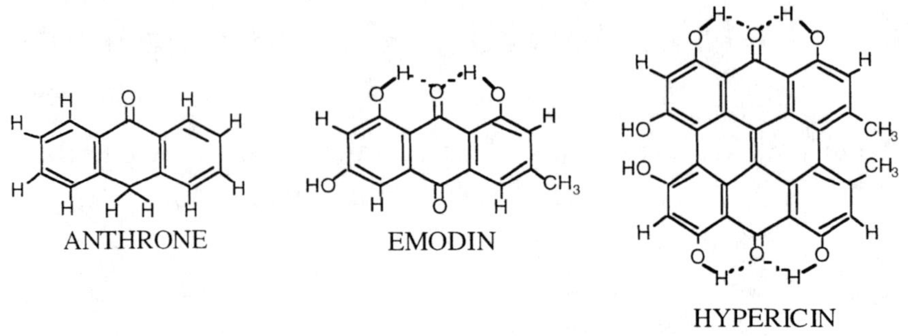

FIGURE 1. Chemical structure of anthrone, emodin and hypericin.

MATERIAL AND METHODS

Powder samples of anthrone and emodin were purchased from Sigma-Aldrich and used as supplied. Neutron inelastic scattering (NIS) spectra of these molecules have been recorded at 20 K on TFXA spectrometer at ISIS neutron spallation source at the Rutheford-Appleton Laboratory (UK). These spectra have been complemented by IR spectra at low (20 K) and room temperatures. FT-Raman spectra of the powder samples of anthrone and emodin have also been obtained at room temperature. The whole set of these spectra have been used in assigning the internal vibrational modes in both low and high wavenumber regions. Theoretical calculations were based on the Hartree-Fock (HF) as well as on the density functional theory (DFT) using B3LYP non-local exchange and correlation functionals. All the computations were performed with the Gaussian94 code on Cray C-94 and C-98 platforms [7]. Geometry optimization as well as harmonic force field calculations were carried out using 6-31G basis functions with non-standard exponents for d-orbital polarization functions, i.e. 0.75 and 0.85 for carbon and oxygen atoms, respectively. This special basis set has been called 6-31G$^{(*)}$. Cartesian Harmonic force constants as output from the quantum mechanical calculations have been described in terms of internal coordinates. Redundant coordinates have then been removed and vibrational wavenumbers as well as potential energy distribution (PED) and NIS intensities have been calculated by means of the home-made program BORNS.

RESULTS AND DISCUSSION

Experimental and calculated (first-order) NIS spectra of anthrone have been reported previously [8-9]. The main features of the vibrational spectra of anthrone have been assigned by means of quantum mechanical calculations at DFT/B3LYP/6-31G$^{(*)}$

level of theory. HF calculations with the same basis sets lead to a poor agreement between the experimental and calculated results. Whatever the level of the theoretical calculations is, the geometry optimization show that anthrone maintains the experimentally observed structure (planar rings with global C_{2v} symmetry). In addition, IR spectra obtained at 295 and 20 K do not reveal any considerable changes in the wavenumber and intensity of vibrational modes. Thus, the vibrational spectra of this molecule is barely influenced by temperature.

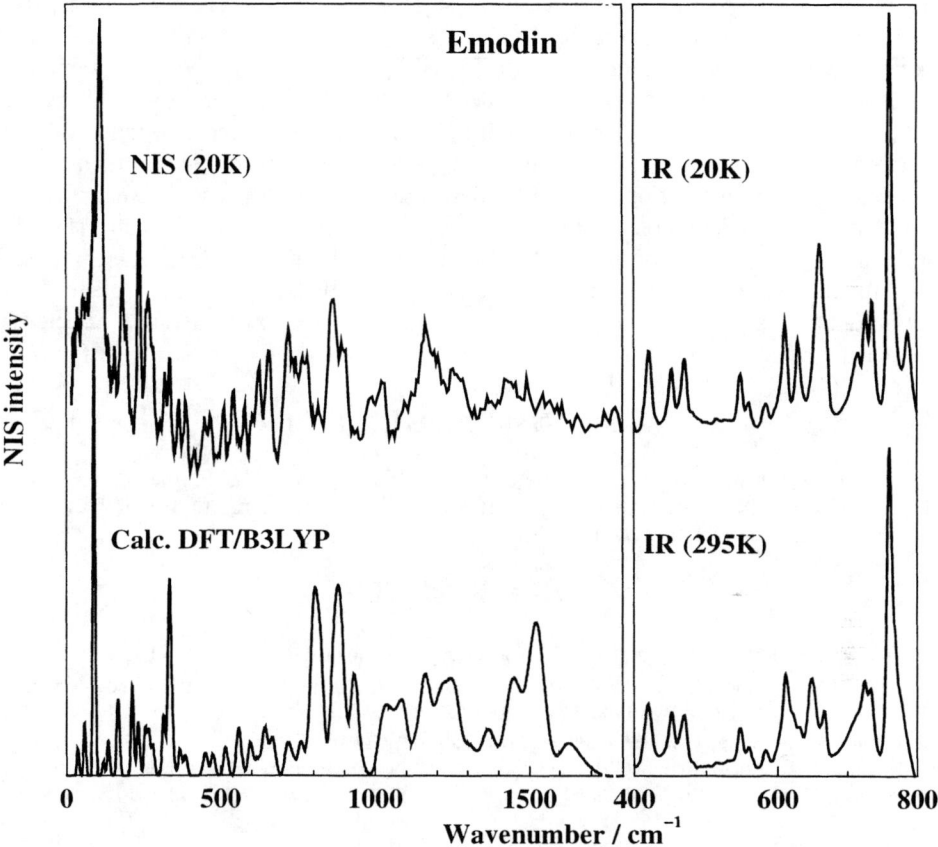

FIGURE 2. Comparison between the experimental and calculated (unscaled force field) NIS spectra of emodin (left). Note that the lattice modes give rise to additional bands below 120 cm^{-1} which are not reproduced by calculations based only on internal modes. IR spectra of powder samples of emodin have been displayed as a function of temperature (right) to show the changes in the 600-700 cm^{-1} spectral region.

As far as emodin is concerned, although the three rings of the molecule remain planar (C_s symmetry), the calculations give us the opportunity to analyze the preferential orientations of the CH_3 and OH groups (Fig. 1). The molecular geometry related to the lowest energy corresponds to one where the two OH groups of the top half of the molecule are involved in intramolecular hydrogen bonds with the central carbonyl group. In the bottom half of emodin, the hydroxyl group as well as one C-H bond of the methyl remain in the plane of the three conjugated rings. Further calculations have also been undertaken in order to analyze the dynamics of the proton in terms of tautomerism, caused by the proton exchange between the adjacent C-O bonds. They show that all these tautomers have energies too high to be considered as major populations at low or room temperatures. The comparison between the experimental and calculated NIS spectra (with the lowest energy conformer), has been displayed in Figure 2 (left). Although no scaling has been applied to the calculated force constants, the main features of the experimental peaks are well reproduced. However, relative intensities are not reproduced in the regions around 1500, 650 and 400 cm^{-1}. These discrepancies may originate from the intermolecular associations in solid phase. The IR spectra shows an influence of temperature on the intramolecular H-bonds in emodin (Fig. 2, right). In fact, the medium bands peaking at 660 and 668 cm^{-1} involving the out-of-plane C-O-H vibrational motions in the top half of the molecule, are considerably affected by increasing the temperature.

ACKNOWLEDGEMENTS

The authors would like to thank the technical staff of the Rutherford-Appletton Laboratory for NIS spectroscopy and IDRIS (CNRS) for computational facilities on Cray C-94 and C-98 supercomputers.

REFERENCES

1. Muruelo, D., Lavie, G. and Lavie, D., *Proc. Natl. Acad. Sci. USA* **85**, 5230-5234 (1988).
2. Lavie, G., Valentine, F., Levin, B., Mazur, Y., Gallo, G., Lavie, D., Weiner D., and Meruelo, D., *Proc. Natl. Acad. Sci USA* **86**, 5963-5967 (1989).
3. Duran, N. and Song, P.-S., *Photochem. Photobiol.* **43**, 677-680 (1986).
4. Hudson, J. B., Lopez-Bozzocci, I. and Towers, G. H. N., *Antiviral Res.* **15**, 101-112 (1991).
5. Sureau, F., Miskovsky, P., Chinsky, L., Turpin, P.Y., *J. Am. Chem. Soc.* **118**, 9484-9487 (1996).
6. Surendra, N. S., *Acta Cryst.* **17**, 851-856 (1964).
7. Frisch, M. J., Trucks, G. W., Schlegel, H. B., Gill, P. M. W., Johnson, B. G., Robb, M. A., Cheeseman, J. R., Keith, T., Petersson, G. A., Montgomery, J. A., Raghavachari, K., Al-Laham, M. A., Zakrzewski, V. G., Ortiz, J. V., Foresman, J. B., Cioslowski, J., Stefanov, B. B., Nanayakkara, A., Challacombe, M., Peng, C. Y., Ayala, P. Y., Chen, W., Wong, M. W., Andres, J. L., Replogle, E. S., Gomperts, R., Martin, R. L., Fox, D. J., Binkley, J. S., Defrees, D. J., Baker, J., Stewart, J. P., Head-Gordon, M., Gonzalez, C., and Pople, J. A., Gaussian, Inc., Pittsburgh PA, 1995.
8. Ulicny, J., Ghomi, M., Jobic, H., Aamouche, A, *J. Mol. Struct.* **410**, 497-501 (1997).
9. Ulicny, J., Ghomi, M., Jobic, H., Miskovsky, P., Berthier, G., *Biological Macromolecular Dynamics*, Eds. Cusack, S., Büttner, H., Ferrand, M., Langan, P., Timmins, P., New York: Adenine Press, 1997, pp. 105-109.

Low Frequency Internal Vibrations of Norbornane and Its Derivatives Studied by IINS and Quantum Chemistry Calculations

K. Holderna-Natkaniec[1], I. Natkaniec[2], V.D. Khavryutchenko[3]

*1 Institute of Physics A. Mickiewicz University. 61-614 Poznan. Poland.
2 Frank Laboratory of Neutron Physics, JINR, 141980 Dubna, Russia,
and H. Niewodniczanski Institute of Nuclear Physics, 31-342 Krakow, Poland.
3 Institute of Surface Chemistry, Nat. Ac. Sci. Ukraine, 252028 Kiev, Ukraine.*

Abstract. The observed and calculated INS vibrational densities of states for globular molecules of norbornane, norborneole and borneole are compared in the frequency range up to 600 cm^{-1}. Inelastic incoherent neutron scattering (IINS) spectra were measured at ca. 20 K on the high resolution NERA spectrometer at the IBR-2 pulsed reactor. The IINS intensities were calculated by semi-empirical quantum chemistry method and the assignments of the low-frequency internal modes were proposed.

Norbornane, C_7H_{12}, is the simplest globular molecule with a rigid framework of polycyclic alkanes. There are two aspects of special interest in the case of its substitution: internal excitations of the molecule and the solid state phase transitions. Analysis of the vibrational spectrum requires the knowledge of the spatial structure and the force field of the molecule. The coordinates of atoms in the molecule were obtained by the quantum chemistry (QC) semi-empirical AM1 method. The norbornyl skeleton with marked the *exo-* and *endo*-positions of oxygen atom, and the calculated structures of the *exo*-norborneole, $C_7H_{11}OH$, and *exo*-borneole, $C_7H_8(CH_3)_3OH$, molecules are presented in Fig.1.

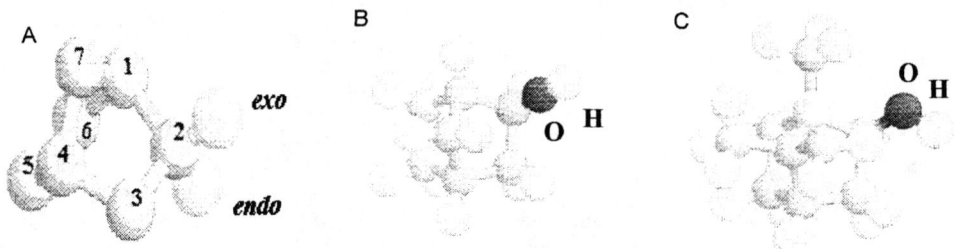

FIGURE 1. The norbornyl skeleton with the notation used for the carbon atoms - (A), the calculated structures of *exo*-norborneole - (B) and *exo*-borneole - (C) molecules.

CP479 *Neutrons and Numerical Methods — N₂M*
edited by M. R. Johnson, G. J. Kearley, and H. G. Büttner
© 1999 The American Institute of Physics 1-56396-838-X/99/$15.00

The bond lengths of C1-C1, C2-C3, C7-C4 and C-H in the norbornane molecule determined by electron diffraction (1) are 1.556, 1.551, 1.559, and 1.115 Å, to the accuracy of (±0.015 Å). The distances mentioned above determined by X-ray diffraction (2) are 1.536, 1.544, 1.536 and 1.076Å. These distances are close to calculated by AM1 method: 1.541, 1.540, 1.550 and 1.115 Å. The bond lengths of C1-C2, C2-C3, C7-C4, C2-O for *exo-* and *endo-*norborneole molecule not known from the experiment were determined by AM1 method as: 1.561, 1.558, 1.546, 1.364 Å and 1.540, 1.540, 1.551, 1.431 Å, respectively. The C-C-C angles in these molecules were less then tetrahedal.

The force field was determined in the Cartesian system of coordinates, as a matrix of the second derivatives of the total energy of the molecule with respect to the atom displacements (3,4). On the basis of the atom coordinates and the force field, the frequencies of particular modes and their intensities in the INS spectrum have been obtained. The frequency calculated for an isolated molecule of norbornane, its qualitative assignment with potential energy distribution (PED), and experimentally obtained frequencies, are collected in Table 1. These frequencies differ by about 10%. The calculated frequencies were fitted to their experimental values by solving the full matrix inverted vibration problem (5). The fitted neutron intensities are shown by bars in Fig. 2. The calculated IINS spectra were obtained by the convolution of these δ-

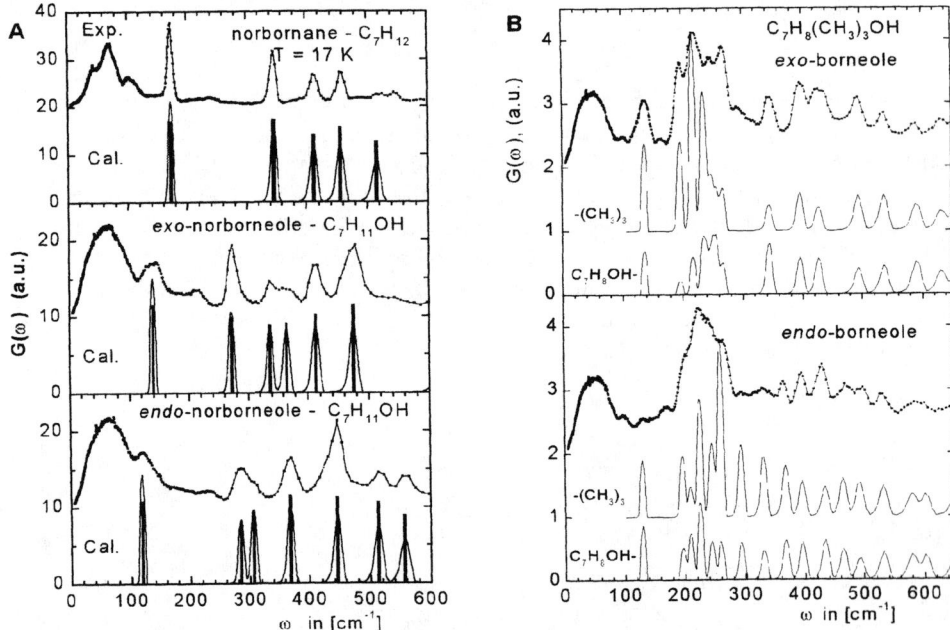

FIGURE 2. **A** - The experimental $G(\omega)$ at 17 K (Exp.) and calculated (Cal.) spectra of the low-frequency internal vibrations for the norbornane, *exo-* and *endo-*norborneole molecules. **B** – comparison of the experimental $G(\omega)$ at 17 K of *exo-* and *endo-*borneole, with the partial $G_H(\omega)$ density of states calculated for the three methyl groups and the C_7H_8OH skeleton of the borneole molecules.

TABLE 1. Comparison of the frequencies and assignments of the internal vibrations of norbornane molecule below 600 cm^{-1}.

IINS experiment at 17 K. [cm^{-1}]	calculated frequencies for isolated molecule [cm^{-1}]	Assignment for isolated molecule and PED %	Fitted frequencies [cm^{-1}]	IINS intensity (after fitting) (a.u.)	Assignments after Fitting procedure and PED %
173	193.	χ[C-C] 63 %	173.	13048.	χ[C-C] 64 %
					χ[C1-C2] 16 % *χ [C3-C4] 16 %* *χ[C4-C7] 16 %* *χ [C6-C1]) 16 %* *χ [C2-C3] 16 %*
347	366.	δ[C-C-C] 25 % χ [C-C] 48 %	346.	13382.	χ [C-C] 35 % δ [C-C-C] 33 % *δ [C2-C1-C6] 14 %.* *δ [C3-C4-C7] 14 %* *χ [C1-C2] 8 %* *χ [C3-C4] 8 %* *χ [C4-C7] 8 %*
411	439	δ[C-C-C6] 75 %	406.	10934.	δ [C-C-H] 38 % *δ [C1-C6-H16] 9%* *δ [C1-C6-H17] 9%* *δ [C4-C7-H8] 9%* *δ [C4-C7-H9] 9%*
456	481.	δ[C-C-C7] 42 %	455.	12129.	δ[C-C-C7] 50 % χ [C-C7] 26 % *χ[C1-C3] 13 %* *χ [C13-C4] 13 %* *δ [C2-C1-C3] 12 %* *δ [C6-C1-C3] 12 %* *δ [C3-C4-C3] 12 %*
516	611	δ[C-C-C] 66 %	516.	9854.	δ [C-C-C]$_{skeleton}$ 62 % *δ[C2-C1-C6] 25 %* *δ [C3-C4-C7] 25 %*

functions with the resolution function of the NERA spectrometer [6], in approximation of the one-phonon neutron scattering cross-section. Finally, the transformation of the calculated and experimental IINS spectra to the G(ω) amplitude weighted phonon density of states was performed. The G(ω) functions are compared in Fig.2.

The fitted low-frequency vibrations of norbornane molecule and qualitative assignment of subsequent modes are presented in Table 1. Particular modes with the PED greater then 10% are marked by *italics*. Each of these low frequency internal vibration is a combination of several torsions: stretching (ν), bending in-plane (δ) and bending out-of-plane (χ). The bands appearing in the low-frequency spectrum of norbornane can be interpreted as corresponding to the breath vibrations of skeleton at 173 cm^{-1}, and subsequently the out-of-plane and in-plane bendings, as in Table 1.

The skeleton deformation modes of the *exo-* and *endo*-norborneole were shifted mainly towards lower frequencies in comparison to those observed for norbornane. For both norborneole molecules the out-of-plane bending χ[CO] at 338 cm^{-1} and 369 cm^{-1}, and in-plane bendingss δ[CCO] at 272, 366 cm^{-1} and 284, 369 cm^{-1}, for *exo-* and *endo-* forms, respectively, were observed.

The G(ω) spectra of norborneole substituted at 1.7.7 by methyl groups i.e. *exo-* and *endo*-borneole, are compared in Fig. 2 B with the results of the calculations. The calculated results are presented as partial spectra of C_7H_8OH skeleton, and three methyl groups. The out-of-plane vibrations of the methyl groups were observed in the range 190 - 400 cm^{-1}. These vibrations are mixed with the skeleton ones. A significant splitting of these modes ascribed to these groups is a consequence of their different molecular surrounding, which is consistent with the results reported by Makita (7) for methyl substituted compounds. It was found that the lines corresponding to the torsional vibration involving any atoms from the hydroxyl group were close to 260 cm^{-1}, and those attributed to the deformational modes δ[CCC-O] are close to 350 - 480 cm^{-1}.

The half-width of the peaks in the experimental spectrum of norbornane, in the region up to 600 cm^{-1}, is almost the same as the resolution function of the spectrometer, i.e. it is close to ca. 15 cm^{-1}. The half-width of the bands observed in this frequency range for the *exo*-norborneole alcohols studied, vary from 20 to 30 cm^{-1}. Broadening of subsequent bands much over the resolving power of the spectrometer as well as the overlapping of the low-frequency internal modes and the lattice ones, especially for *endo*-borneole is shown in Fig. 2. The wide bands of the low frequency internal vibrations and structureless lattice vibration density of states below ca. 100 cm^{-1}, lead to the conclusion about that the orientational disorder of molecules in the low-temperature phase of the studied norborneole alcohols is frozen. The neutron diffraction pattern for *endo*-norborneole and *endo*-borneole does not indicate the phase transition in the temperature range from 17 to 300 K (8). Therefore, the orientational glass state of these alcohols in the low-temperatures may be concluded.

REFERENCES

1. Chiang, J.F., Wilcox, C.F., Bauer, S.H., *J.Am. Chem.Soc.*, **90**(12) 3149-3157 (1968).
2. de Carneiro, J.W., Seidl, P.R., *J.Mol.Struct*, **152,** 281-291 (1987).
3. Smith, J.C., Karplus, K., *J.Amer.Chem.Soc*, **114**(3), 801- 812 (1992).
4. Dewar, M.S.J., *J.Molec.Struct*, **43,**135-137 (1978).
5. Khavryutchenko, V.D., *The COSPECO complex of programs to computed vibrational spectroscopy*, Institute of Surface Chemistry, Nat. Ac. of Science Ukraine, Kiev, 1990.
6. Natkaniec, I, Bragin, S.I., Brankowski, J., Mayer, J., *Proc. ICANS XII*, Abingdon 1993, RAL Report 94-025, Vol. I., p. 89-96.
7. Makita, K., Kagayama, A., Sarto, Y., Uno, T., *Spectrochim.Acta*, **34A** (4) 909 - (1978).
8. Holderna-Natkaniec, K., Natkaniec, I., *Proc. XI Polish Conf. Molec.Cryst*, Gdansk 1998.

Neutron Spectrometry and Numerical Simulations of Low-Frequency Internal Vibrations in Solid Xylenes

I. Natkaniec[1], K. Holderna-Natkaniec[2], J. Kalus[3], V. D. Khavryutchenko[4]

*1 - Frank Laboratory of Neutron Physics, JINR, 141980 Dubna, Russia,
and H. Niewodniczanski Institute of Nuclear Physics, 31-342 Krakow, Poland.
2 - Institute of Physics, A. Mickiewicz University, 61-614 Poznan, Poland.
3 - Institute of Physics, University of Bayreuth, 95440 Bayreuth, Germany.
4 - Institute of Surface Chemistry, Ukrainian Academy of Science, 252028 Kiev, Ukraine.*

Abstract. Vibrational densities of states of solid xylenes were determined from the inelastic neutron scattering spectra measured on the NERA spectrometer at the IBR-2 pulsed reactor. These spectra were used to test the semi-empirical quantum-chemistry calculations of internal vibrations of xylene molecules with differently deuterated sub-units. Rotations of methyl groups were found to be strongly affected by intermolecular interactions in the crystals and mixed with phenyl ring deformations.

A molecule of xylene, $C_6H_4(CH_3)_2$, contains a phenyl ring and two methyl groups at the *para-*, *metha-* and *ortho-* positions, as shown in Fig. 1. Determination of a vibrational spectrum requires the knowledge of spatial structure and force-field of a molecule. The calculations were performed for an isolated molecule of the compounds studied. The atom coordinates in the molecule were obtained by the AM1 semi-empirical quantum chemistry (QC) method (1) and the Hartree-Fock method with the 6-31G basis set of the GAUSSIAN 94 program (2).

FIGURE 1. The calculated structures of the *para-*, *metha-* and *ortho*-xylene molecules.

The structures of the molecules as well as the frequencies of normal modes calculated by these two methods do not differ significantly. Using the molecular structure and the force field matrix as calculated by the AM1 method, the inverse spectroscopic problem was solved to fit the calculated frequencies of isolated molecules to these experimentally observed in crystals (3).

The force field was determined in the Cartesian system of coordinates, as a matrix of second derivatives of the total energy of the molecule with respect to atom displacements. On the basis of the atom coordinates and the force field matrix, the frequencies of particular modes - ω_j, and their intensities in the amplitude weighted vibrational density of states - $G(\omega)$, were obtained. The contribution of the *n-th* atom involved in the *j-th* vibrational mode to the $G(\omega)$, was calculated as proportional to the scattering cross-section - σ_n, and the displacement - $a_n(\omega_j)$,

$$G(\omega) = \Sigma_n ([\sigma_n]^2/M_n) \Sigma_j [a_n(\omega_j)]^2 \delta(\omega-\omega_j). \tag{1}$$

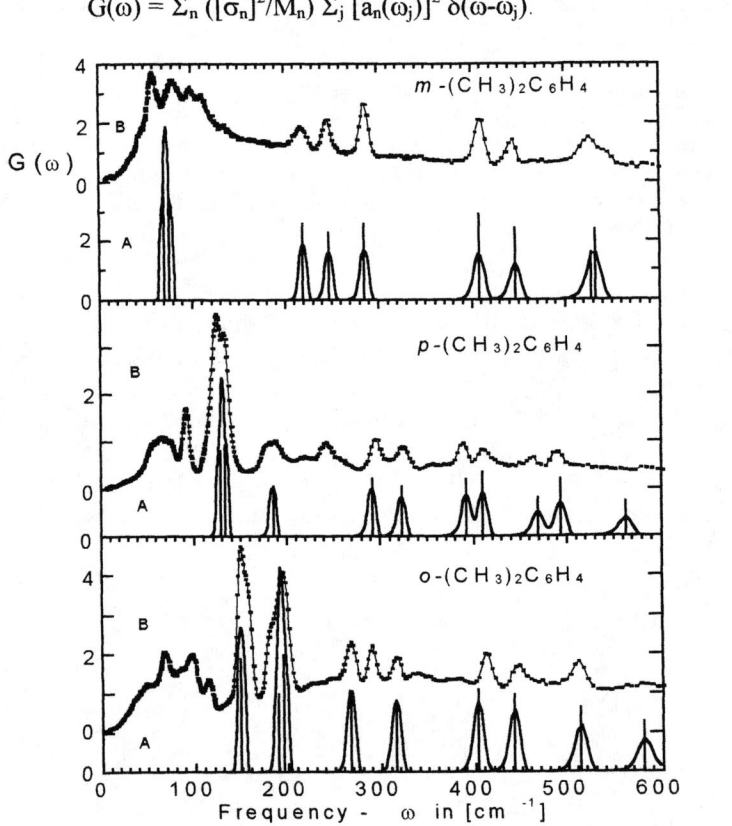

FIGURE 2. A comparison of the neutron scattering intensities (A - bars) calculated according to the formula (1) and the simulated spectra spectra (A - solid lines) of internal vibrations up to 600 cm^{-1}, with the $G(\omega)$ experimental spectra (B) of solid xylenes at 10 K.

The calculated IINS spectra were obtained in the approximation of one-phonon scattering process by convolution of the neutron intensities calculated according to formula (1), with the resolution function of the NERA spectrometer (4). The inverse transformation of the calculated and experimental IINS spectra into the $G(\omega)$ spectra presented in Fig. 2, was made according to the one-phonon scattering cross-section formula, without solving the deconvolution problem.

The frequencies of the torsional modes of methyl groups for the isolated molecules of m-, p- and o-xylenes, were calculated by the AM1 method as ca. 60 cm^{-1} for the former two and about 200 cm^{-1} for latter. These modes are experimentally observed at about 60-80 cm^{-1} for solid m-xylene, and their frequencies are located in the ranges: 120-140 cm^{-1} and 150-200 cm^{-1} for solid p- and o-xylenes, respectively. It means that inter-molecular forces in solid xylenes play a significant role in the rotational dynamics of the methyl groups. The calculated frequencies of other internal modes below 600 cm^{-1} differ by about 10% from the experimental ones. These low frequency deformations of the phenyl ring are mixed with the methyl groups rotations. Neutron intensities and simulated $G(\omega)$ spectra presented in Fig. 2, were computed by fitting the calculated frequencies to the experimental ones. Due to mixed character of the low frequency modes, the fitting procedure was accomplished by solving the inverted vibration problem with the full force matrix (3).

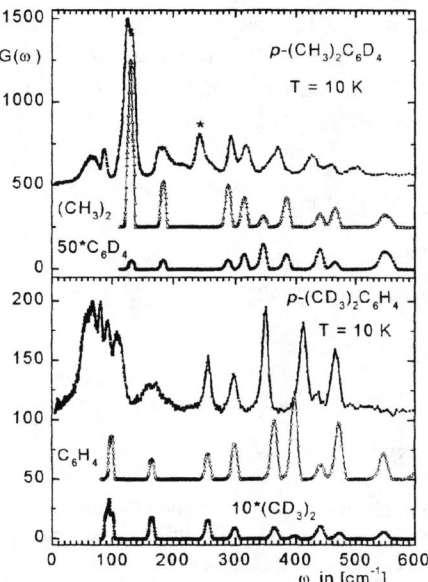
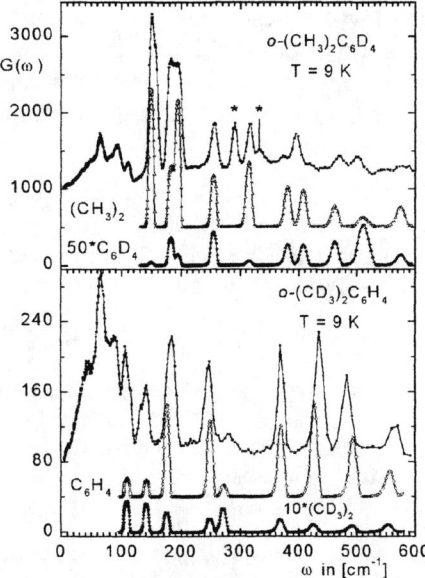

FIGURE 3. A comparison the $G(\omega)$ spectra of partially deuterated p- and o-xylene molecules with the calculated partial $G(\omega)$ spectra of the two methyl groups -(CH$_3$)$_2$ and the phenyl ring C$_6$H$_4$- of the corresponding molecules. The partial spectra of deuterated sub-units -(CD$_3$)$_2$ and C$_6$D$_4$- should be multiplied by the factors of 10 and 50, respectively, to see them in the scale of the protonated sub-units. The second harmonics of the methyl group torsions are marked by asterisk.

The IINS spectra of deuterated xylene molecules were also measured in order to test the correctness of the fitted force constants matrices and assignments of the low-frequency modes shown in Fig. 2. The spectra of xylene molecules with the deuterated C_6D_4- and -$(CD_3)_2$ sub-units were calculated using the force constants matrices fitted to the experimental frequencies of $C_6H_4(CH_3)_2$ molecules. The experimental and calculated spectra of p- and o-xylenes are presented in Fig. 3. In the case of each partially deuterated sample, the deuterated sub-units almost do not contribute to the IINS spectra of the protonated sub-units. The experimental $G(\omega)$ in this case corresponds to the partial weighted density of states of the protonated sub-units.

The most intensive bands in the $G(\omega)$ spectra of xylenes correspond to the torsions of CH_3 groups. These modes are quite weak in the partial spectra of the phenyl ring, as one can see in Fig. 3. The low frequency deformations of the phenyl ring in xylene molecules are quite intensive in the partial spectra of CH_3 groups. The torsional frequencies of deuterated methyl groups are not shifted by the factor $1/\sqrt{2}$, because of they are mixed with the lattice vibrations. These modes should be taken into account in the lattice dynamics calculations of the xylene crystals. Such calculations based on the experimental crystal structure, which is known for p-xylene, are presented in the references (5) and (6).

Good agreement between the calculated and experimental frequencies as well as neutron intensities for partially deuterated molecules of xylenes confirm the assignments of the low-frequency internal vibrations implied by the computation of the simulated spectra as presented in Fig.2. Our recent results illustrate the usefulness of the neutron spectroscopy and relatively fast semi-empirical QC calculations for the investigation of the low-frequency methyl group vibrations, which are very poorly seen by optical spectroscopy methods.

ACKNOWLEDGMENTS

The authors' thanks to Eng. S.I. Bragin and Dr. Ch. Munch for their help with the experiment and Dr. A. Pawlukojc for the QC calculations made with the help of the GAUSSIAN 94.

REFERENCES

1. Philipenko, A.T., Zayetz, A.V., Khavryutchenko, V.D., Falndysh, E.P.,
   Zh. Struk. Khimii, **28**, 155-157 (1987).
2. Frisch, M.J., Trucks, G.W., Schlegel, H.B. et all., *GAUSSIAN 94*,
   Gaussian Inc., Pittsburgh PA, 1995.
3. Khavryutchenko, V.D., Khavryutchenko, A.V., Khavryutchenko, Al.V., "Quantum Chemistry Toolset for Simulation of Structure and Vibration Spectra of Molecules and Clusters," presented at the Workshop: Neutrons and Numerical Methods, ILL, Grenoble 9-12 Dec. 1998.
4. Natkaniec, I., Bragin, S.I., Brankowski, J., Mayer, J., "Multicrystal Inverted Geometry Spectrometer NERA-PR at the IBR-2 Pulsed Reactor," *in Proceedings of the ICANS XII Meeting, Abingdon 1993*, RAL Report 94-025, Vol. I., p. 89-96 (1994).
5. Kalus, J., Monkenbusch, M., Natkaniec, I., Prager, M., Wolfrum, J., Worlen F.,
   Mol. Cryst. Liq. Cryst., **268**, 1-20 (1995).
6. Natkaniec, I, Kalus, J., Griessl, W., Holderna-Natkaniec, K., *Physica B*, **234-236**, 104-105 (1997).

Molecular dynamics simulation of inelastic neutron scattering spectra of librational modes of water molecules in a layered aluminophosphate

A.J. Ramirez-Cuesta[a,b], P.C.H. Mitchell[a], S.F. Parker[c], A.P. Wilkinson[d] and P. Mark Rodger[a]

[a] *Department of Chemistry, University of Reading, Reading RG6 6AD, UK,* [b] *Departamento de Fisica, Universidad Nacional de San Luis, 5700 San Luis, Argentina,* [c] *Rutherford Appleton Laboratory, Chilton, Didcot, Oxon OX11 0QX UK and* [d] *School of Chemistry and Biochemistry, Georgia Institute of Technology, Atlanta, Georgia 30332-0400, USA.*

Abstract. Molecular dynamics (MD) is well established as a method for computing static and dynamical properties of solids. In this work we present a comparison of the experimental and calculated structures and vibrational spectra calculated using MD and the experimental spectrum from inelastic neutron scattering of a templated and hydrated layered aluminophosphate. We discuss the water librational modes between 300-800 cm^{-1} where the assignment of peaks in the experimental spectrum is not clear. We have identified the librational modes from the trajectories of the water molecules. The various librational modes were decomposed into their components about the main rotational axis of the water molecules, and the different weight of each libration in the final spectrum was thereby established. From this analysis we were able to correlate three librational modes with water molecules in three different environments.

INTRODUCTION

Fabrication of materials with chiral nano-environments is a developing area of nano-technology. Such materials will be applied in, for example, the separation of racemic mixtures into enantiomerically pure compuonds in biology and medicine. Chiral cobalt complexes have been used as structure directing templates in the synthesis of layered aluminophosphates (ALPOs) (1). The ALPO DL-$[Co(en)_3]Al_3(PO_4)_4 \cdot 3H_2O^{\dagger}$ consists of ALPO layers with $[Co(en)_3]^{3+}$ cations and water molecules in the interlayer region (2). We are studying the interactions between the template, the water molecules and the ALPO layers, in particular whether the interlayer water molecules have a structural role. Water plays an important role in biological self-assembly, for example by stabilising biopolymer conformations through hydrogen bonding (3). According to our computational modelling of layered ALPO's water may similarly mediate the

---

† en: 1,2-diaminoethane; $H_2NCH_2CH_2NH_2$.

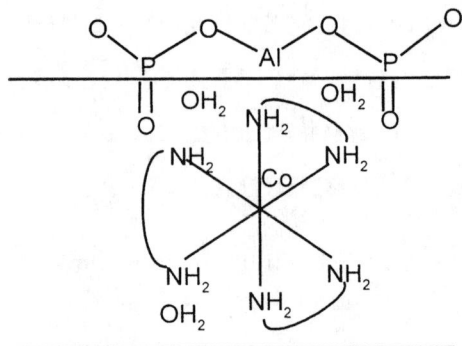

FIGURE 1. Templated and hydrated layered aluminophosphate.

interaction between the templating $[Co(en)_3]^{3+}$ cation and the ALPO layer. X-ray crystallography was able to locate the oxygen atoms of the interlayer water molecules in our templated ALPO but the positions of the hydrogen atoms could only be inferred. In a preliminary communication we reported how inelastic neutron scattering (INS) and molecular dynamics (MD) simulations enabled us to locate the hydrogen atoms and to describe the hydrogen bonding of the water molecules (4). A detailed interpretation of the observed vibrational spectra based on neutron scattering data alone is not possible. However, information can be obtained by performing molecular dynamics simulations and calculating neutron scattering spectra from the resulting atomic trajectories (5). This approach has been used in characterising dynamics in molecular crystals (6), methylene chloride (7) and proteins (8). Here we present a more complete account of our simulations.

THE SIMULATION SCHEME

The structure of the template and the layered ALPO are shown in FIGURE 1; note the water molecules between the template and the ALPO layer. Clearly there are hydrogen bonding possibilities involving hydrogens of the water molecules, oxygen of the phosphate groups, nitrogens of the template, and hydrogens of the template with phosphate and water oxygens.

The simulation procedure is summarised in FIGURE 2. The X-ray structure provides an initial set of atomic coordinates. The force field which describes the interatomic and intermolecular interactions is derived from our earlier work, water molecules and template molecules are considered as rigid unit without internal degrees of freedom (9). The molecular dynamics program DL_POLY (10) was used to generate simulated atom trajectories and coordinates which enabled us to test how far a simulated structure reproduced the X-ray structure at finite temperature, see table 1. The use of

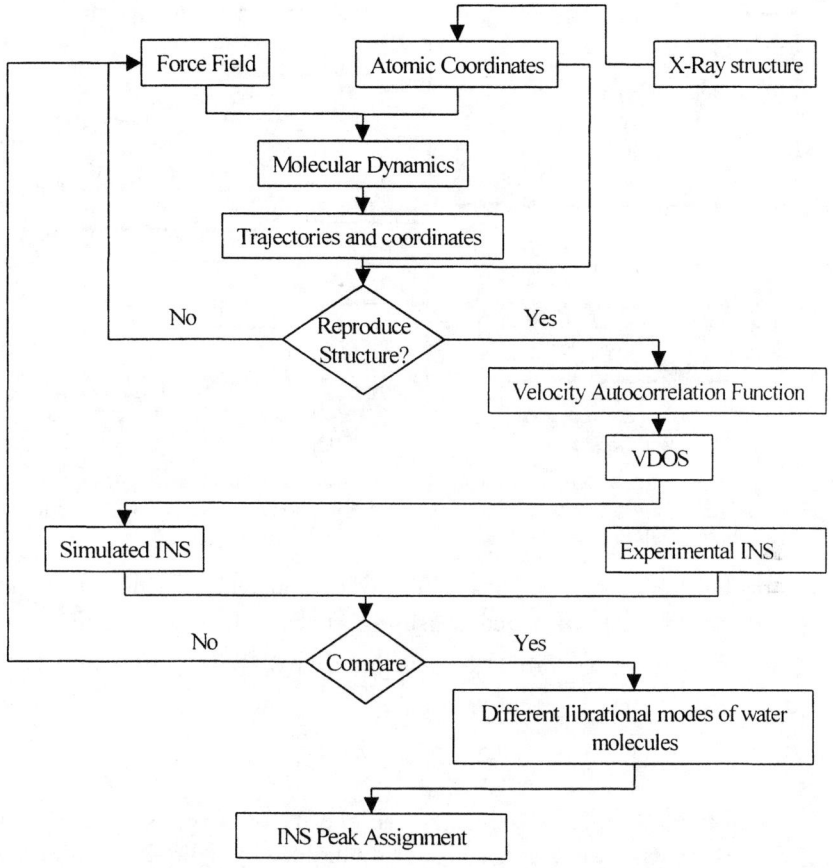

FIGURE 2. Flowchart of the simulation procedure and comparison with experiment.

MD in this calculation allows us to study the structural parameters of the solid at a finite temperature, thus providing a better comparison with experiment. We were able to reproduce the experimental structure to within a root means square deviation (rmsd) of 2%. We stress that this agreement was obtained with MD in the NST ensemble, and as such involved no symmetry constraints at all, not even constraining the shape or volume of the unit cell. Simulated INS and infrared spectra calculated from the velocity autocorrelation function and the vibrational density of states were compared with the experimental spectrum. This comparison provided a further test of the force field, more sensitive than the purely structural comparison. Following this procedure we were able to choose the force field which best reproduced the vibrational spectra (modified from that of van Beest, van Santen potential (9,11)). Those parts of the simulation which refer to calculation of the INS spectra are described below.

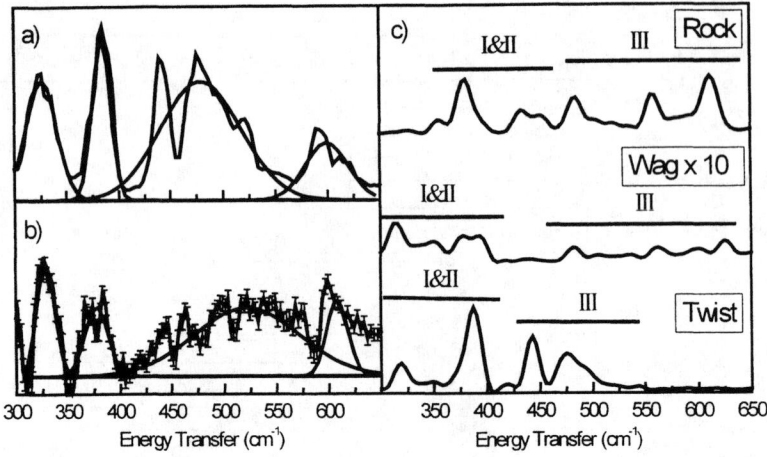

FIGURE 3. a) Calculated and b) experimental difference spectra in the 300-650 cm^{-1} region. Gaussians are there to show the four major peaks c) Librational modes of water molecules decomposed according with their motion.

We focus on the librational motions of the water molecules, which occur in the region 250–1000 cm^{-1}. The calculated density of states, $G(\omega)$, was obtained as the Fourier transform of the auto-correlation function $\langle \mathbf{v}_\alpha(0) \cdot \mathbf{v}_\alpha(t) \rangle$ of velocities $\mathbf{v}_\alpha(t)$ for atoms α

$$G(\omega) = \frac{1}{N} \sum_\alpha \frac{1}{2\pi} \int_{-\infty}^{\infty} e^{-i\omega t} \langle \mathbf{v}_\alpha(0) \cdot \mathbf{v}_\alpha(t) \rangle dt$$

where ω is frequency (12). Experimentally, the density of states is weighted by the scattering cross sections of the atoms and so is dominated by hydrogen atom motions; thus calculations of $G(\omega)$ are reported for α = hydrogen.

The simulation procedure involves a relaxation at low temperatures; a series of molecular dynamics calculations at various temperatures up to 300K; then quenching the system to 20K. For the calculation of the trajectories the simulations were performed under constant number of particles, volume and energy, NVE ensemble, during this process.

EXPERIMENTAL

INS spectra of the templated ALPO, DL-[Co(en)$_3$]Al$_3$(PO$_4$)$_4$.3H$_2$O, and the templating complex, [Co(en)$_3$]Cl$_3$, both hydrated (2H$_2$O) and anhydrous, were recorded on the TFXA spectrometer at the Rutherford Appleton Laboratory ISIS facility (13). The compounds (10g) in aluminium foil sachets were examined at 20 K over energy transfers in the range 16-4000 cm^{-1}.

The main features of these spectra arise from H atoms, associated with either ethylenediamine ligands or water molecules. Bands arising from ethylenediamine, were found to be coincident in all three spectra. Thus the experimental spectrum for water in the hydrated AlPO was obtained by subtracting the $[Co(en)_3]Cl_3$ spectrum from the templated ALPO, d,l-$[Co(en)_3]Al_3(PO_4)_4.3H_2O$ (with intensities normalised to the ethylenediamine peak at 320 cm^{-1}). This spectrum is presented in figure 3 along with a smoothed spectrum obtained by Gaussian deconvolution of the experimental difference spectrum.

RESULTS

To obtain a detailed assignment of the water frequencies, the calculated spectrum has been decomposed into contributions from librations about the three principal molecular axes (defined in FIGURE 4). The density of states is dominated by librations around the x- and z-axes. rocking and twisting, with less than 10% of the intensity in the y-axis libration, wagging. All three librational modes exhibit a number of separate peaks, indicating distinct environments for the water molecules. The first two peaks in the water librations spectra, labelled I & II, are the low frequency motion of the molecules and these modes are due to water molecules less tightly bonded; these two water molecules share a large cage. For the high energy librations, peaks labelled III, there is one water molecule per cage, but it can be bounded with three different NH groups from a neighbouring template, producing then three different peaks corresponding to different metastable orientations for the water molecule. In consequence the broad peak exhibit a more complex structure in the experiment as well as in the simulation.

CONCLUSIONS

Our simulation of the structure of DL-$[Co(en)_3]Al_3(PO_4)_4.3H_2O$ has been compared with experiment. The force field used was tested and it correctly reproduces the structure of the solid from X-ray crystallography. It also has been used to predict the INS spectra of the compound and it successfully compared with experiment. A further analysis was made in order to test the possible motions of water in the ALPO. We found the relative contribution of each librational mode to the overall spectrum and we identified the different environments that are responsible for the various peaks in the experimental spectrum.

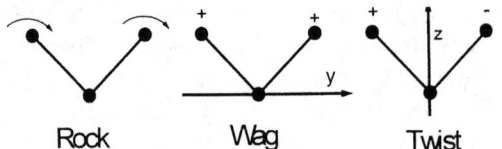

FIGURE 4 Three principal rotational / librational modes for water

ACKNOWLEDGEMENTS

We thank EPSRC for financial support under grant GR/K90463 and for allocation of computing time on the J90 Cray supercomputer at the CCLRC Atlas Centre (PCHM) and EPSRC and Rutherford Appleton Laboratory for beam time. We thank J. Kruger from School of Chemistry and Biochemistry, Atlanta for the provision of the ALPO sample.

REFERENCES

1. M.E.Davis and R.F.Lobo, *Chem. Mater.* **4**, 756 (1992). D.A. Bruce, A.P. Wilkinson, M.G. White, A. Bertrand, *J. Chem. Soc. Chem. Commun.* 2059 (1995).
2. K. Morgan, G. Gainsford and N. Milestone, *J. Chem. Soc. Chem. Commun.*, 425 (1995).
3. L.J. Barbour, G.W. Orr and J.L. Atwood, *Nature*, **393**, 671 (1998).
4. A.J. Ramirez-Cuesta, P.C.H. Mitchell, A.P. Wilkinson, S.F. Parker, and P.M. Rodger; *J. Chem. Soc. Chem. Commun.*, 2653 (1998).
5. A.J. Dianoux, J.L. Sauvajol, G.R. Kneller and J.C. Smith, *Journal of non-crystalline Solids*, 472, (1994).
6. G.R. Kneller, W. Doster, M. Settles, S. Cussack and J.C. Smith, *J. Chem. Phys.*, **97**, 8864 (1992)
7. G.R. Kneller and A. Geiger, Molec. Phys. **70**, 465 (1990).
8. J.C. Smith, S. Cusack, B. Tidor and M. Karplus, J. Chem. Phys. **93**, 2974 (1990).
9. A.J. Ramirez-Cuesta, P.C.H. Mitchell and P.M. Rodger, *J. Chem. Soc., Faraday Trans.*, **94**, 2249 (1998).
10. T.R. Forester and W. Smith, DL_POLY User Manual, CCLRC, Daresbury Laboratory, version 2.0, 1995.
11. B.W.H. Van Beest, G.J. Kramer and R.A. van Santen, *Phys. Rev. Lett.*, **64**, 1955 (1990).
12. A.J. Dianoux, G.R. Kneller, J.L. Sauvajol and J.C. Smith, *J. Chem. Phys.* **99**, 5586 (1993), L. van Hove, Phys. Rev. **95**, 249 (1954). L. van Hove, Physica, **24**, 404 (1958)
13. J. Penfold and J. Tomkinson, Rutherford Appleton Laboratory Report, RAL-86-019, 1986.

On the Origin of the Distribution of Potential Barriers for Methyl Group Dynamics in Glassy Polymers: Neutron Scattering & MD-simulations

F. Alvarez*, A. Alegría*, J. Colmenero*,
T. M. Nicholson[†], G. R. Davies[†]

*Departamento de Física de Materiales, Universidad del País Vasco
Facultad de Química, Apto. 1072, 20080 San Sebastián, Spain.
[†] IRC in Polymer Science and Technology, University of Leeds, Leeds LS2 9JT, U.K.

Abstract. We have carried out molecular dynamics simulations of methyl group torsional librations in glassy polyisoprene at 150 K using the Insight and Discover programs from MSI and the Polymer Consortium Force Field. The model system used was built using the MSI Amorphous Cell Builder. During the dynamics runs, the position and velocity of the atoms as well as the dihedral angle of each of the methyl groups were recorded at 10 fs intervals. The results obtained support the threefold approximation for the single particle methyl group potential. The density of states for methyl group torsional librations, calculated from the time evolution of the dihedral angles, agrees quite well with neutron scattering results and shows a broad feature reflecting a broad distribution of potentials barriers. Performing similar simulations, but under "phantom-chain" conditions, we conclude that the width of this distribution is mainly controlled by the non-bonded interactions.

INTRODUCTION

Recent neutron scattering experiments of quantum and classical methyl group dynamics in glassy polymers have shown the main role played by distributions of potential barriers for methyl group rotation (1-3). The results from both, quantum and classical temperature regimes, of different polymers can consistently be described in the framework of the rotation rate distribution model (RRDM) which assumes a pure single particle threefold potential ($V(\varphi) = V_3(1 - \cos 3\varphi) / 2$) for methyl group rotation together with a Gaussian distribution of potential barriers $G(V_3)$. Although in general these distributions should be associated to the structural disorder of any amorphous system, their microscopic origin cannot be directly inferred from the experiments. Due to the fast time scale of methyl group dynamics, molecular dynamics (MD) simulations seem to be a suitable tool to gain insight into this problem. In this work, we have carried out a MD simulation of methyl group torsional librations in glassy polyisoprene (PI) at 150 K. The main interest of this work is to check the threefold approximation in a realistic simulation and to investigate the main microscopic parameters controlling the width of the distribution of potential barriers.

MODEL AND METHOD

Simulations were carried out using the Insight (Insight II 4.0.0.P version) and Discover-3 programs from MSI (4) using the Polymer Consortium Force Field (PCFF). The model system used in these simulations was built using the MSI Amorphous Cell Builder. The polymer system that we wanted to mimic was the glassy polyisoprene (PI) used in previous inelastic neutron scattering measurements (5). The PI microstructure modelized was 90 % of 1,4-PI and 10 % of 3,4-PI. All the 1,4-PI units were chosen to be cis. A cubic cell containing 3 PI-chains of 10 monomeric units each was constructed at a density of 0.96 gcm^{-3} and a temperature of 298 K, leading to a cell dimension of 15.5 Å. Periodic boundary conditions were employed in order to model a bulk system. The range of nonbonded interactions considered between atoms was smaller than half the cell side so no atom interacts with an image of itself. Five separate cells were independently built to ensure that a range of conformations of the amorphous polymer was explored. The different cells constructed were first subjected to an energy minimization procedure in order to equilibrate them. It is worth emphasizing that the temperature initially considered (298 K) is about 100 K higher than the glass transition temperature of the real PI samples, allowing to equilibrate the system in a short time scale. After this procedure, the system was "quenched" to the glassy state at a temperature of 150 K similar to the temperature at which the reported neutron scattering measurements were carried out. The density of the system was ajusted to that of this temperature leading to a new dimension of the cubic cell of 15.2 Å. After a new equilibration procedure, MD simulations at this temperature were run on each of the constructed cells. Initially, the atoms of the system are assigned velocities from a Boltzman distribution at the corresponding temperature. Then, Newton's laws of motion are solved numerically in 1 fs time steps using the Velocity Verlet algorithm. A NVT emsemble was used with the Velocity Scaling algorithm to keep the temperature constant. The molecular dynamics were run for 160 ps. MD simulations were also run under "phantom-chain" conditions, i.e., by switching off the nonbonded interactions. In this case the dynamics were run only during 5 ps.

During the dynamics runs, the dihedral angle $\varphi(t)$ corresponding to each of the methyl group protons was recorded at 10 fs steps. It is worthy of remark that this angle defines the rotation of a given methyl group proton in the plane perpendicular to the bond joining the methyl group to the rest of the molecule. In order to calculate the neutron scattering static structure fartor S(Q) of the simulated system, the coordinates of all the atoms were also recorded at each step. The calculated S(Q) was found to agree quite well with the experimental data measured with spin polarization analysis in the Q-range from 0.5 Å to 2.5 Å by means of D7 spectrometer (ILL, Grenoble).

RESULTS AND DISCUSSION

At the chosen low temperature, the dihedral angle $\varphi(t)$ is, in most cases, a fluctuating function around one of the three states approximately given by the values of φ of 0° or ±120°. However, there are still some methyl groups that are undergoing a significant number of transitions between these states during the simulation runs. Therefore, MD results seem to support the threefold jump model which is usually assumed for analyzing neutron scattering data of methyl group classical rotation. In order to carefully check this question, we have calculated the dihedral angle distribution averaged over all methyl group protons and trajectories. The results obtained are shown in Fig. 1, which clearly indicates a threefold distribution. On the other hand, we have also calculated the distribution of the jump-dihedral angle averaged over all the methyl group protons jumping during the simulation runs. This distribution only shows a narrow peak centered on 120°, thereby also in good agreement with the threefold approach. It is worth mentioning that previously reported simulations of poly(methyl methacrylate) (6) also agree with this conclusion.

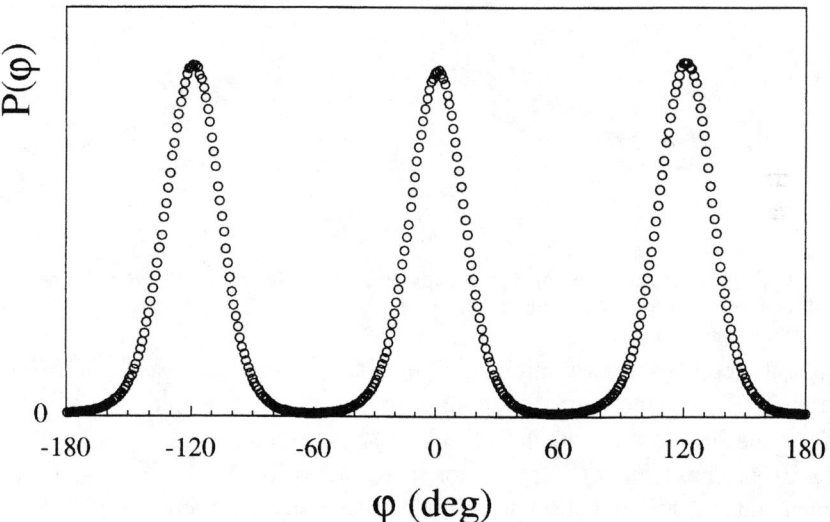

FIGURE 1. Dihedral angle distribution averaged over all methyl group protons and trajectories

From the MD simulation results, the density of torsional libration states for methyl groups can be obtained in terms of the angular velocity autocorrelation function as:

$$g(E) \propto \int_0^\infty e^{-iEt} \langle \dot{\varphi}(0)\dot{\varphi}(t) \rangle \, dt \qquad (1)$$

The results obtained are shown in Fig. 2 in comparison with neutron scattering results. It is worth emphasizing that g(E) cannot be directly obtained from the inelastic neutron scattering data which always measure the total density of vibrational states. However, in a first approximation, g(E) can experimentally be calculated from the difference between the hydrogen weighted density of states corresponding to a PI sample with the methyl group protonated and the main chain deuterated and that of another PI sample with the methyl group deuterated and the main chain protonated (see Fig. 7 of ref. 5). This is what is represented in Fig. 2. The agreement of the calculated g(E) from MD simulations with the experimental results indicates that both, the average methyl rotational barrier and the distribution of potential barriers, are well reproduced with the used force field.

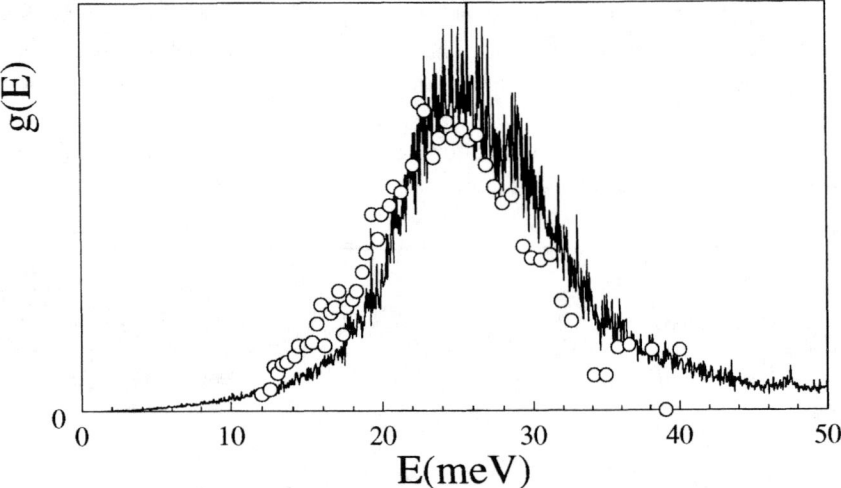

FIGURE 2. Density of states for methyl group torsional librations in PI from MD simulations (solid line) and from neutron data (empty points)

Once the threefold approximation has been tested, we can easily transform the calculated g(E) into the corresponding distribution of threefold potential barriers $G(V_3)$. In the framework of the threefold approximation and assuming that the measured (or simulated) density of torsional states only corresponds to the first librational state, a numerical relationship between the librational energy E and the potential barrier V_3 can easily be stablished from the solutions of the corresponding Schrödinger equation. This numerical correlation can be well approximated by a power law which in the range considered here ($V_3 < 3000$ K) reads as:

$E(meV) = 0.476 (V_3)^{0.55}$, where V_3 is in units of Kelvin. According to this expression, $G(V_3)$ is given by $G(V_3) = g(E) \, dE / dV_3$. The results obtained are shown in Fig. 3. As it was expected from the broad torsional libration $g(E)$ peak, the obtained distribution function $G(V_3)$ also shows a broad feature with a maximum centered in 1375 K and a half width at half maximum (HWHM) of about 550 K. These are typical characteristics of glassy polymers (2). Now, we can take advantage of the possibilities of MD-simulations to get some insight into the microscopic origin of this distribution. To do this, we have repeated all the simulations and calculations above described but in "phantom-chain" conditions, i.e., by switching off the nonbonded interactions at the beginning of the MD runs. The results obtained for $G(V_3)$ in this case are also shown in Fig. 3. As can be seen, the maximum of $G(V_3)$ is now shifted towards the low enery range, but the main effect is a strong narrowing of the distribution (HWHM $\approx$ 180 K) which indicates that the nonbonded interactions are the main parameter controlling the width of this distribution.

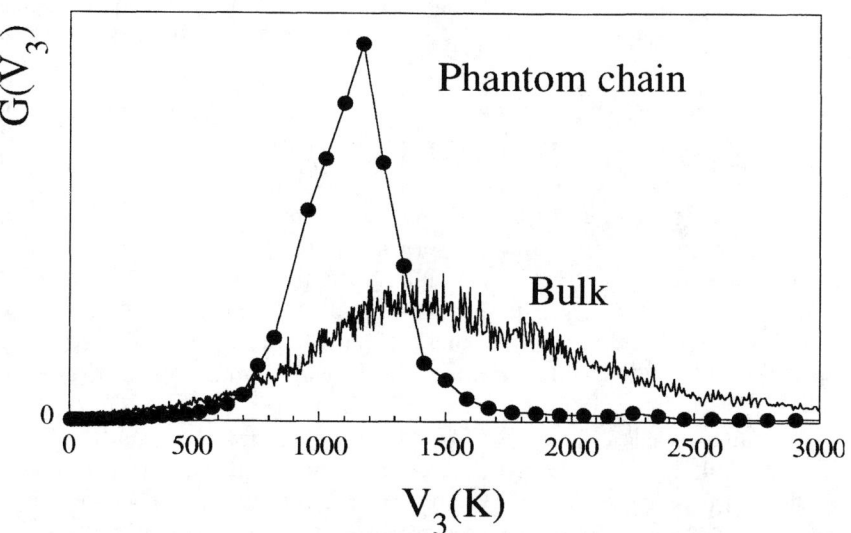

FIGURE 3. Distribution of threefold potential barriers for methyl group dynamics in PI from MD simulations

REFERENCES

1. Colmenero, J., Mukhopadhyay, R., Alegría, A., Frick, B., Phys. Rev. Lett. **80**, 2350-2353 (1998)
2. Mukhopadhyay, R., Alegría, A., Colmenero, J., Frick, B., Macromolecules **31**, 3985-3993 (1998)
3. Moreno, A., Alegría, A., Colmenero, J., Frick, B., Phys. Rev. B (in press)
4. Molecular Simulations Inc., San Diego
5. Frick, B., Fetters, L. J., Macromolecules **27**, 974-980 (1994)
6. Nicholson, T. M., Davies, G. R., Macromolecules **30**, 5501-5505 (1997)

Density functional theory for the calculations of the rotational potentials of methyl groups

Béatrice Nicolai[†] and Gordon J. Kearley[†]

[†]*Institut Laue langevin, BP156, 38042 Grenoble, FRANCE*

Abstract. The rotational dynamics of methyl groups in 2,6 dimethylpyrazine and in cobalt and nickel acetates have been studied by neutron scattering and a combination of ab initio and empirical calculations. These results and the precision of the method are compared to those obtained by DFT calculations in each case. It has been shown that DFT gave excellent agreement with observables for organic molecules. We determined the significant parameter in the evaluation of the potential in ionic metal-containing systems.

INTRODUCTION

Methyl group tunneling frequencies are very sensitive probes of the local rotational potential [1]. Calculations in a series of crystalline compounds demonstrate that a combined *ab initio* and molecular mechanics method (MM) to obtain rotational potentials gives reasonable results [2–5]. The remaining discrepencies between observed and calculated values in some systems may arise from the limitations of MM calculations, based on empirical pair-potentials and partial charges centered on atomic positions.

Density functional methods (DFT) provide an accurate alternative for calculations of rotational potentials in the crystalline unit cell. The results of the DFT calculations are in very good agreement with experimental data for the organic compound: 2,6-dimethylpyrazine and are promising for ionic compounds such as the metal-acetates $(M(CH_3COO)_2.4H_2O$ with M=Co, Ni). The advantages and limitations of DFT will be discussed and compared to those of the combined *ab initio* and MM method in these systems.

PRINCIPLES

According to quantum mechanics, the energy of a molecule can be obtained by solution of the Schrödinger equation. The case of a hindered symmetric rigid rotor, with a fixed rotational axis, can be solved using the single-particle approach, the simplest approach to methyl rotation. In this theory, there are two kinds of excitations: the tunnelling transitions ν and the librational transitions.

The almost exponential variation of ν as a function of the rotational potential makes tunnelling spectroscopy a very sensitive probe of the molecular crystalpotentials. We evaluated the rotational potential in the crystal with 2 methods and compared it to experimental data.

2,6 DIMETHYLPYRAZINE

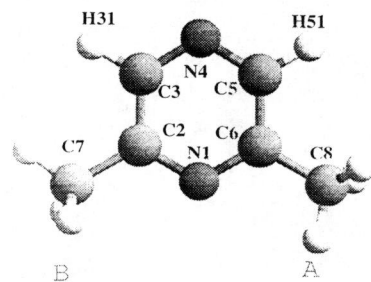

FIGURE 1. Molecular geometry determined by neutron diffraction at 2K.

The low-temperature structure of 2,6 dimethylpyrazine has been solved at the Laboratoire Leon Brillouin (Saclay, France) by single crystal neutron diffraction. This compound contains two crystallographically inequivalent methyl groups attached to the aromatic pyrazine ring [6]. In the crystal at low temperature, these two methyl groups are in a staggered conformation (fig 1). One C-H bond of methyl group (C8) points towards the N1-atom (conformation A) whereas the C-H methyl group of methyl group (C7) points towards the H31-atom (conformation B).

Each of the methyl groups experiences different potentials due to environmental differences. The inelastic neutron scattering spectrum exhibits two tunnelling lines at 20 and 29 μeV and two libration lines at 8 and 10 meV, one due of each of the crystallographycally-distinct methyl groups. Assignment of these spectroscopic observations to the inequivalent methyl groups has been made directly by inspection of the tunnelling-excitation intensity as a function of the momentum-transfer direction in an oriented single crystal [4].

We have investigated the specific dynamics of these methyl groups by calculation of the rotational potential from the crystal structure, using GAMESS-UK *ab-initio* program [7] and Cerius^2 molecular mechanics software [8]. It was found that none of the 3 contributions (internal, van der Waals an Coulomb) were negligible. The rotational potentials were overestimated by a factor of 1.5 and 2. However, the experimental conformation is correctly predicted and it arises essentially from the intermolecular contributions [4].

We performed a DFT calculation on a unit cell of 2,6 dimethylpyrazine with the commercial programm CASTEP [9]. The agreement with the observables are remarkable. The staggered experimental conformation is correctly predicted and the tunnelling frequencies are underestimated by a factor of 2 which corresponds to an error on the barrier height of only 10 percent.

We compared the electrostatic potential energy surfaces obtained with *ab initio* on the isolated molecule and with partial charges centered on atomic positions, as used in MM methods. In *ab initio*, there are additional minima corresponding to lone pairs of nitrogen atoms and to the aromatic ring (fig 2 and 3), these minima are absent in the empirical methods. Because electrostatic are long range interactions and because methyl groups are surrounded by aromatic rings and N atoms, these discrepencies may have a non negligible effect on the estimation of the intermolecular barrier.

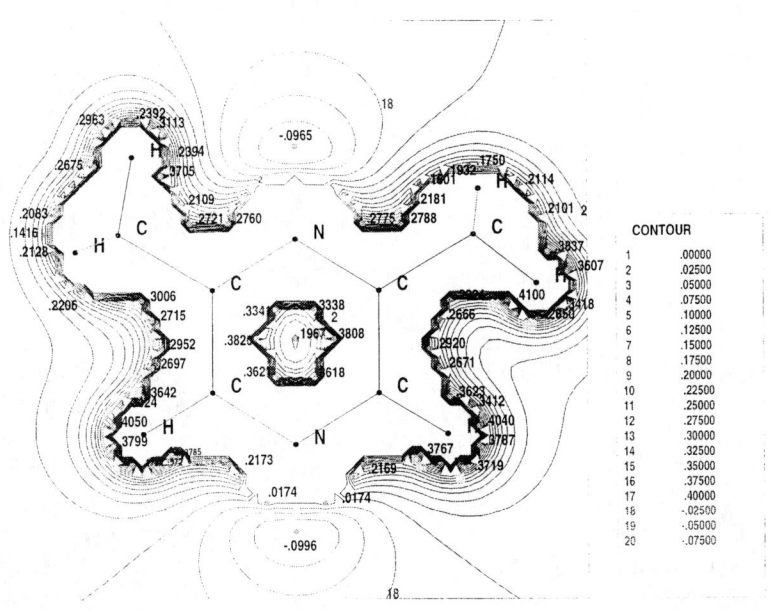

FIGURE 2. Electrostatic potentials calculated for the isolated molecule by ab initio. The energies are in Hartrees

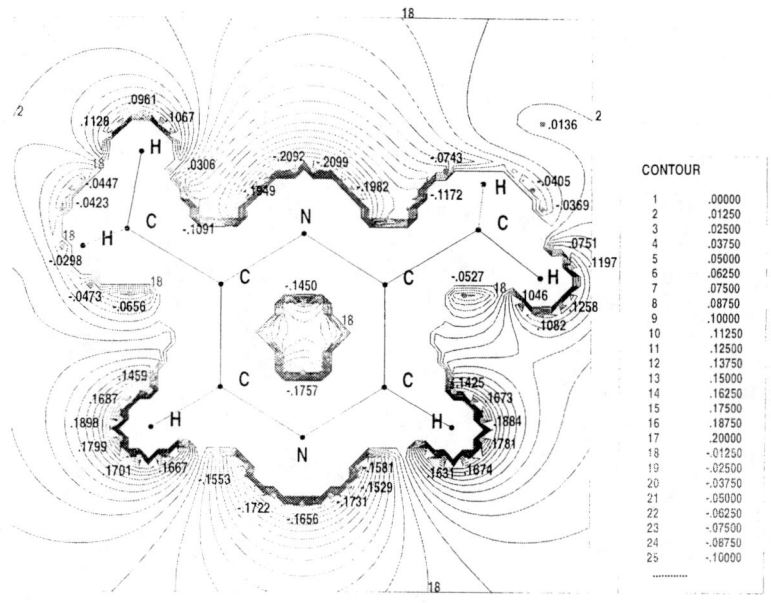

FIGURE 3. Electrostatic potentials calculated for the isolated molecule by MM method. The energies are in Hartrees

COBALT AND NICKEL ACETATES

We have determined the low-temperature crystalline structure of the cobalt and nickel acetates tetrahydrate by neutron powder diffraction [5]. They crystallize in the space group $P2_1/c$. They are found to be almost isostructural. We performed inelastic neutron scattering measurements: the 2 acetates show a single tunnelling transition at about 1.5 μeV on IN10 [5].

We performed the calculations of the rotational potential from the crystal structure [5], using the combination of *ab-initio* and molecular mechanics. It was found that the Coulomb interaction is the main contribution to the rotational potential in these ionic compounds. A good agreement with the observables (the methyl group orientation and the tunnelling and librational transitions) is obtained if the partial charges are the potential derived charges calculated by *ab initio* and recalculated at each step of the rotation of the methyl group. [5]

In the DFT calculations in these metal-containing scompounds, we demonstrated that the barrier height seems is almost independant of the method (LDA or GGA)

and the cutoff, in contrast to the total energy of the system. The total energy of a system is calculated in DFT by numerical summation on k points, chosen as a function of the symmetry of the crystal. In organic crystals, a limited number of k points is required. In these ionic metal-containing systems, the electronic density is very different from the density of an organic system. These metal-containing systems require an increase of the number of k points for an accurate description of the total energy in the crystal. The increase of the number of k points leads effectively to a better agreement with observables (fig 4). We note that for these metal containing systems it is essential to impose the wavefunction initialisation for each calculation.

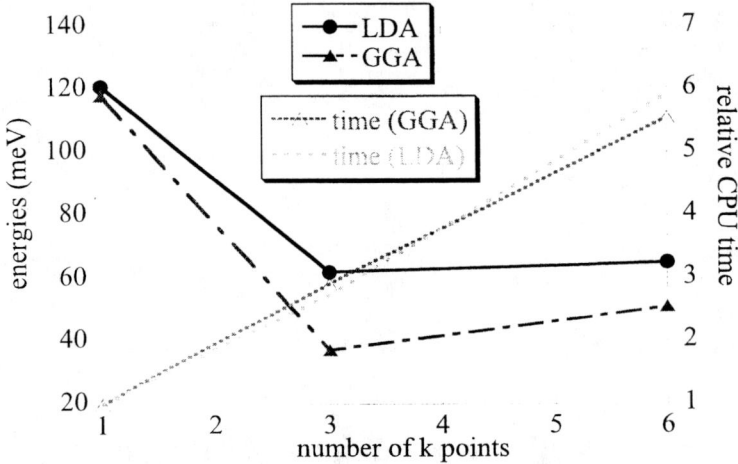

FIGURE 4. Calculated barrier height in cobalt acetate, as a function of the number of k points and the method. The experimental barrier height is 48meV.

CONCLUSION

The combination of *ab initio* and MM method gives reasonable results but it appears limited because it doesn't take into account properly the electrostatic effects of lone pairs or aromatic rings in a system.

DFT is an alternative method, which allows us to perform *ab initio* calculations in the crystal. It gives very good results on organic compounds, but seems limited by the chemical composition of the system (by the presence of transition metal in the acetates).

REFERENCES

1. W. Press, *Single particle rotations in crystals* Springer tracts in modern physics, **92** (1981).
2. M. Neumann and M.R. Johnson, *J. Chem. Phys.* **107**, 1725 (1997).
3. P.Schiebel, G.J. Kearley and M.R. Johnson, *J. Chem. Phys.* **108**, 2375 (1998).
4. B. Nicolai, E. Kaiser, F. Fillaux, G.J. Kearley, A. Cousson and W. Paulus *Chem. Phys.* **26**, 1 (1998).
5. B. Nicolai, G.J. Kearley, M.R. Johnson, F. Fillaux and E. Suard *J. Chem Phys*,**109** 9062(1998)
6. E. Kaiser *thesis*, Paris XI (1997)
7. M.F. Guest, J. Kendrick, J.H. van Lenthe, K. Schoeffel and P. Sherwood *Generalized Atomic and Molecular Electronic Structure System* Daresbury laboratory, UK.
8. Cerius2 *BIOSYM. Molecular simulations* 8685 Scarton road san diego, California 92121-3753 (1996).
9. M.C. Payne, M.P. Teter and D.C. Alain *Reviews of modern physics* **64**, 1045 (1992).

Rotation-precession and rotor-rotor coupling in 4-methyl-pyridine

M. A. Neumann[1,2], M. Plazanet[2], M. R. Johnson[1]

[1]Institut Laue-Langevin, BP156, 38042 Grenoble Cedex 9, France

[2]Laboratoire de Spectrométrie Physique, Université Joseph Fourier Grenoble I, 38420 St. Martin d'Hères, France

Abstract. The low temperature rotational dynamics of methyl groups in 4-methyl-pyridine is explained in terms of rotation-precession and rotor-rotor coupling. Initial estimates of the precession angle and the rotational potentials are obtained from molecular mechanics calculations. Experimental spectra are calculated from these potentials by numerical solution of Schrödinger's equation for clusters of coupled rotors embedded into a greater ensemble of rotors treated in the mean field approximation. The precession angle and the rotational potentials are adjusted to reproduce new high resolution INS spectra of protonated 4-methyl-pyridine measured at well defined spin temperatures. Excellent agreement with the experimental data is obtained for fully protonated 4-methyl-pyridine. Studying methyl group dynamics in 4MP offers the opportunity to measure potential energy changes on experimentally determined trajectories, thus providing valuable information that can be used to test the reliability of ab initio calculations and empirical force fields for the solid state.

I INTRODUCTION

The investigation of methyl group quantum dynamics by inelastic neutron scattering (INS) started in 1975 with the observation of rotational excitations in crystalline 4-methyl-pyridine (4MP) [1]. Since then it has been widely recognised that methyl groups can be used as probes of their chemical environment, and a recent study has used methyl group dynamics in a variety of molecular crystals to test methods of potential energy calculation [2]. However, the true nature of the low temperature dynamics in 4MP is still a subject of controversy and increasingly exotic quatum mechanical models [3] [4] have been used to explain the growing quantity of experimental data. In this paper we provide an new interpretation of the experimental observations based on molecular mechanics calculations and the numerical solution of Schrödinger's equation for many coupled one-dimensional rotors.

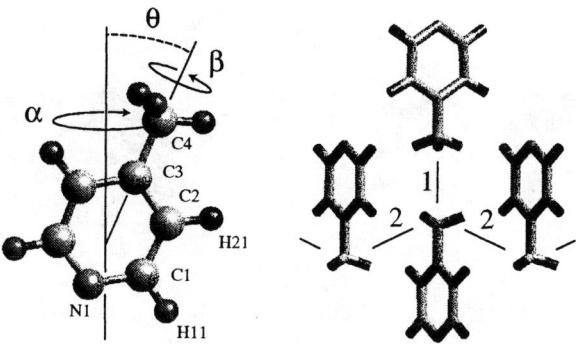

FIGURE 1. Rotation-precession of a single 4MP molecule and the most important coupling types.

II THEORY

The coupled quantum motion of a system of n rigid methyl groups in a rigid environment is described by a Hamilton operator comprising a kinetic energy term for each methyl group and a potential energy terms for the ensemble of methyl groups.

$$H = \sum_{i=1}^{n} -B_i \frac{\partial^2}{\partial \varphi_i^2} H + \sum_{i=1}^{n} V_i^s(\varphi_i) + \sum_{i,j=1, i<j}^{n} V_{i,j}^p(\varphi_i, \varphi_j) \qquad (1)$$

φ_i is the rotation angle of methyl group i, I_i is the corresponding moment of inertia and $B_i = \hbar^2/2I_i$ is called the rotational constant. The numerical solution of Schrödinger's equation for such a Hamilonian and the calculation of neutron scattering matrix elements have been discussed in detail elsewhere [8]. The total wavefunction consists of a rotational part and a nuclear spin part of appropriate symmetry. The symmetry of the nuclear spin functions of the individual methyl groups can be classified by symmetry labels A, E_a and E_b. The interconversion between nuclear spin species is a slow process at low temperature. For a comparison of experimental and calculated intensities, the experimental population of the energy levels has to been known. This has been achieved by first waiting for complete spin relaxation at a temperature were spin relaxation is reasonably fast and rapid cooling to a temperature of 1.5 K where spin relaxation is sufficiently slow to be neglected on the time scale of the neutron scattering experiment.

In 4MP the rotation of the methyl group is coupled to a precession of the whole molecule (see fig. 1) [7], resulting in an important increase of the moment of inertia. The coupled system can move almost freely along a one dimensional trajectory in the multidimensional coordinate space. The coordinate φ_i now descibes the motion along the rotation-precession trajectory, and once the trajectory is know, its parametrisation and the rotational constant can be calculated from eq. 2 , where $R_{j,i}$ and $m_{j,i}$ are the mass and the position of atom j in molecule i:

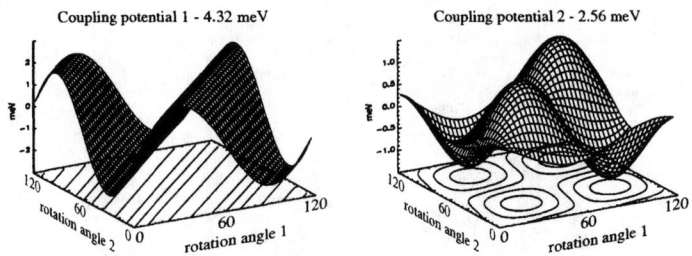

FIGURE 2. The most important coupling potentials.

$$B_i = \hbar^2/2 \sum m_{j,i} (\frac{dR_{j,i}}{d\varphi_i})^2 = const \tag{2}$$

An additional complication arises from the fact that a torque is needed that tilts the molecule back to its mean orientation in order to keep the tumbling molecule on the rotation-precession trajectory. This torque is provided by the forces between the molecule and its environment. As a consequence, with inceasing rotational energy the amplitude of precession will increase and the rotational constant will decrease. Neglecting this effect leads to a systematic overestimation of higher energy levels.

III MOLECULAR MECHANICS CALCULATIONS

The space group of 4MP at 10 K is I $4_1/a$ with 8 equivalent molecules in the unit cell, the rotation axis of the methyl group being a C_2 symmetry axis [5]. Intramolecular contributions to the potentials energy and atomic point charges are determined by Hartree-Fock calculations at the MP2/6-31G** level. Intermolecular Van der Waals interactions are taken into accound using the exponential-6 functional form with parameters from the Universal Force Field [6]. As an estimate for the rotation-precession trajectory the minimum energy path was calculated. The minimum energy path overestimates the precession amplitude and predicts a strong eliptic motion of the methyl group carbon atom in disagreement with crystallographic data. Therefore a rotation-precession trajectory was used that is similar to the minimum energy path but in better agreement with experiment. The molecule precesses around a fix point which is located at 1/3 of the distance from atom N1 to atom C3 and the methyl group carbon atom C4 moves on a circle defined by the precession angle θ, which is the only adjustable parameter of the model. Rotational single particle potentials and coupling potentials (CP) have been calculated for a rotation-precession trajectory with $\theta = 0.9178°$. The dominant coupling type is the coupling of coaxial pairs of methyl groups (CP 1). Another coupling potential connects methyl groups in infinite chains (CP 2) (see fig 1 and fig. 2). The calculated single particle potential had sixfold symmetry and an amplitude of about 1.8 meV.

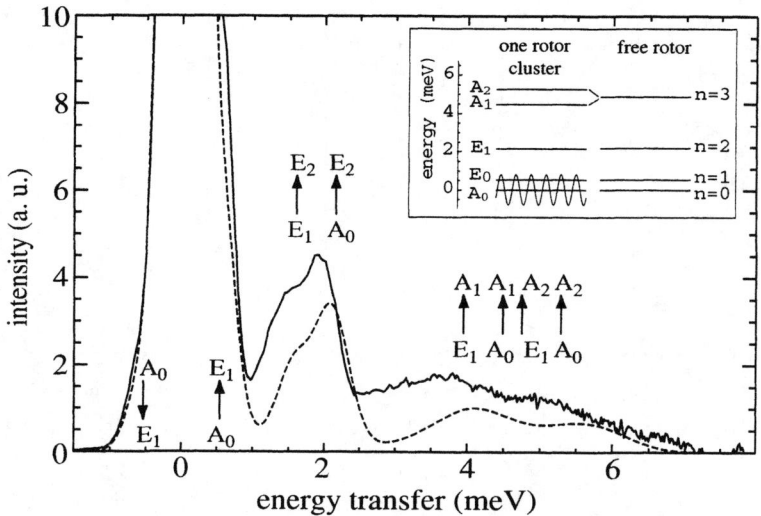

FIGURE 3. Experimental and calculated IN5 spectrum at a spin temperature of 5 K

IV COMPARISON WITH EXPERIMENT

The precession angle θ, the amplitude and the phase factor of the single particle potential and scale factors for the coupling potentials have been adjusted to the experimental data. The precession angle is found to be $\theta=0.8993°$ corresponding to a precession radius of the methyl group carbon atom of 0.053 Å and a rotational constant of 540.3 μeV. Thus the effective rotational constant is 18.5% smaller than the one for a methyl group with fixed rotation axis (663.1 μeV). The adjusted single particle potential had an amplitude of about 0.8 meV and the coupling potentials 1 and 2 were scaled by a factor of 0.638 and 0.66, respectively.

The inelastic neutron scattering spectrum of 4MP-h_7 has been measured on IN5 at the ILL in the energy range from -1.5 to 8 meV. The experimental spectrum obtained at a lattice temperature of 1.5 K and a spin temperature of 5 K is shown in fig. 3 together with a spectrum calculated for a cluster of 6 coupled rotors. The calculated transitions have been convoluted with a gaussian resolution function of 0.25 meV HWHM below 2.5 meV and with a gaussian of 0.5 meV HWHM above 2.5 meV. The broadening of the peaks above 2.5 meV is presumably a consequence of coupling with a great number of additional rotors which could not be taken into account explicitly. The nature of the main features of the spectrum can be understood by comparison with the energy level scheme of a 1 rotor cluster. The intensity differences between the calculated and the experimental spectrum can be attributed to multiple scattering effects. The calculated spectra seems to be stretched to higher energies, because the increase of the precession amplitude with rotational energy has been neglected in the calculation.

The tunnel spectrum has been measured on IN10 and corresponds to transitions

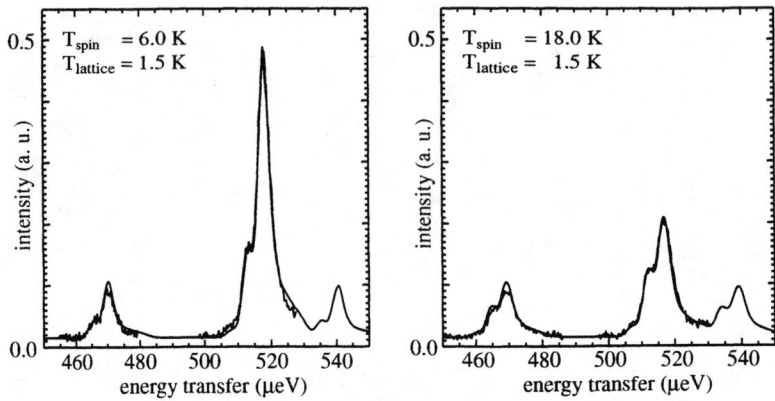

FIGURE 4. Observed and calculated tunnel spectra at different spin temperatures.

from the lowest A state to the lowest E state in the framework of the single particle model. The structure of the tunnel spectrum stems from the coupling with neighbouring rotors. This coupling depends on the nuclear spin species of the neighbouring rotors, and the random distribution of spin species creates a multitude of different environments around a central methyl group. The comparison between experimental data and a calculation for a cluster of 12 coupled rotors is presented in fig. 4. The experimental spectra have been measured at different spin temperatures and the calculated spectra have been convoluted with the asymmetric experimental resolution function. The calculations accurately reproduce the intensity changes as a function of spin temperature.

REFERENCES

1. B. Alefeld, A. Kollmar, B. A. Dasannacharya, J. Chem. Phys. **63**, 4495 (1975)
2. M. Neumann and M.R Johnson, J. Chem. Phys. **107**, 1725 (1997)
3. C. J. Carlile et al, Chem. Phys **134**, 437 (1989)
4. F. Fillaux, C. J. Carlile, G. J. Kearley, Phys. Rev. B **44**, 12280 (1991)
5. Thesis of E. Kaiser Morris, Université Paris XI, 1997
6. A. K. Rappe et al., J. Am. Chem. Soc **114**, 10024 (1992)
7. M. Neumann and G. J. Kearley, Chem. Phys. **215**, 253-260 (1997)
8. M. A. Neumann et al., J. Chem. Phys. **109** (1998), to be published

Tribromomesitylene structure at 14 K Methyl conformation and tunnelling

F. Boudjada*, J. Meinnel*, A. Cousson§ , W. Paulus*§ , M.Mani* # and M. Sanquer*

*Université de Rennes 1, Campus de Beaulieu, 35042 Rennes-Cedex
§ L.L.B. CEA-CNRS, 91191 Gif-sur-Yvette Cedex -
# Université d'El-Jadida Marocco

Abstract. The structure of TBM was solved at 14 K by neutron scattering. Raman, infra-red and neutron spectra were recorded in the range 20-2000 cm-1. The calculation of the internal frequencies done by MOPAC- PM3 allows their assignment. We show also that the quantum coherence between the protons of one methyl group explains the wide delocalization observed.

STRUCTURE OF THE TRIBROMOMESITYLENE AT 14 K

Using the four-circle diffractometer 5C2 at LLB, Saclay, we have established the structure of the tribromomesitylene (TBM) at 14 K. The cell is triclinic, space group $P\bar{1}$ with two molecules per cell related by a center of symmetry (1). We measured 7121 reflections, the refinement on 191 parameters gave a reliability factor R = 4.0 %. The quality of the data allows to say that the molecules have a quasi-perfect C_{3h} symmetry, the angles inside the aromatic ring are 124.2(2)° facing a bromine and 115.8 (2)° facing a methyl group, each methyl group has a C-H bond of 1.072(3) Å eclipsed in the molecular plane, while the two staggered C-H bonds with 1.083(3) Å are slightly longer. In the ring the mean carbon-carbon bond is 1.393(2) Å facing the eclipsed C-H and 1.403(2) on the other side, the ring-methyl C-Cm = 1.496(2) Å and C-Br = 1.895(2) Å , there is clearly a steric repulsion between the bromine and the eclipsed C-H. Figure 1 gives the probability density map in the plane of the three protons of one methyl, we see three elongated ellipsoids, to each proton correspond three different thermal coefficients : the largest of all is tangential U_{11} = 0.105 Å^2, while radially we have U_{22} = 0.014 Å^2 and perpendicularly, corresponding roughly to the C-H stretching U_{33} = 0,023 Å^2, we will explain further the origin of U_{11} and the " banana" shape deformation.

When the temperature rises, progressively the shape of the "protons cloud" is modified (1) and the complex motion of the molecular skeleton may explain the occurrence of four instead of three maxima of protonic density at 293 K.

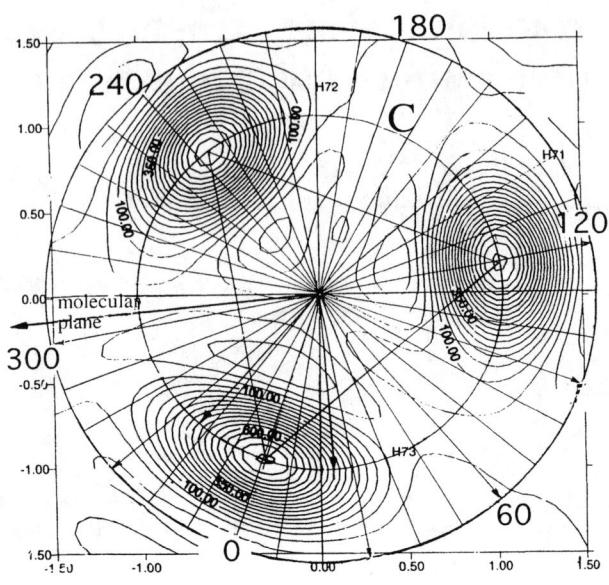

FIGURE 1. Protonic probability density : Fourier-difference map in the protons plane

In parallel to this experiment, we have done quantum chemistry calculations with the semi-empirical program MOPAC-PM3. After optimization, we have found values which are very close to the experimental ones, for the cycle angles : 123.5° and 116.5°, for the C-H bonds 1.0893 and 1.0899 Å, for C-methyl 1.481 Å and C-Br 1.869 Å. From the molecular orbitals, we can know the electronic charges located on each atom, they are very small : for C and Br, -0.131 e and +0.023 e, for C and Cm 0.00 e and -0.089 e, for the eclipsed and the staggered C-H bonds, the charges on the hydrogens are respectively +0,062 e and +0.068 e. The electrons are flowing mainly from hydrogens to carbons atoms and there is a small but significant reinforcement of the double bond character for the C-C bond facing the eclipsed C-H bond.

SPECTROSCOPY : INTERNAL VIBRATIONS, LATTICE MODES AND METHYL GROUPS EXCITATIONS

This quantum chemistry program MOPAC allows (2) to calculate the internal modes of the TBM, the tables 2 and 3 give the results of the FORCE sub-program in the PM3 approximation. The calculated frequencies, the symmetry of the modes are given in tables 1, 2 and 3, they may be compared with the experimental data which were obtained by : Raman diffusion at 4 K (20-3500 cm-1), infra-red at

293 K (600-4000 cm-1), far infra-red at 4 K (20-400 cm-1) and neutron scattering at 4 K. Neutron data were measured on the triple axis spectrometer 4F2 at LLB (20-130 cm^{-1}) and in a wider range (25-3200 cm^{-1}) with the spectrometer TFXA at RAL. The lattice modes have been calculated (2)(3) using a Buckingham potential (exp, r^{-6})

Assignment of observed frequencies in TBM (units cm^{-1})

TABLE 1. Lattice modes

neutron	I.R.	Raman	Buck.pot	Assign.
		23	22	Ra
27		28	28	Rb
32				
35		39	37	Tc
44			44	Rc+Tb
48	45		49	Tb+Rc
	53	51	41	Rc anti
58	58		67	Rb anti
72	72		77	Ra anti
102	110		91	Ta

TABLE 2. TBM internal modes

neutron	I.R.	Raman	PM3	Assign.
70		73	73	a"2
74	72		78	e"
141	147			
154	157	157	133	e'
164	167		169	a"2
230		235	252	a'1
?		279	261	a'2
296	296	295	291	e'
333		338?		e"
381	380?	379	383	e'
452	425?	457		
527	510			
562		568	660	a'1
585			592	a'2
645	646			a"2
698		693		e"
741?				
768		770?	722	e'
	906			
951	954	953	1156	e'
1144		1137	1198	a'1
1167				
		1269		
		1300?		
1514		1510?	1583	a'1
1531	1542	1544	1606	e'
1662		1620?		
1758			1359	a'2
1845				
2032				
		2195	2193	
			3159	

TABLE 3.a. Methyl quantum rotor

neutron	I.R.	Raman	from Pot	Assign.
48	45		49-52	E_{01}
70	72	73	69-71	E_{01}
74	72		73-75	E_{01}
141	147		120-132	E_{02}
154	157	157	134-143	E_{02}
164	167		152-160	E_{02}

TABLE 3.b. Methyl group modes

neutron	I.R.	Raman	PM3	Assign.
205	209	206	210	torsion
1020	1019	1014?	1034	rocking
1078		1058	1053	rocking
1376	1377	1379	1374	bend.s.
1432	1448	1440	1429	bend. a
2900?		2919	3083	stretch.s
2960	2941	2952		comb.
3040	3022	3022	3168	stretch.a

(3), the values given are relative to the center of the Brillouin zone as the peaks measured in Raman or IR. In table 1, the maxima in neutron scattering correspond to maxima in the density of states, slight differences in the positions may be due to the dispersion versus q in the Brillouin zone.

If we study the motion of atoms other than hydrogen, the molecule may be treated as having the symmetry D_{3h} giving the assignment of Table 2. The agreement is generally very good, we have also done a calculus of the frequencies with a valence force field which confirms this assignment (2). In the table 3.b, the methyl internal modes were classified separatly as the streching, bending, rocking and torsional modes, they are found in the usual range around 2950, 1400, 1020 and 200 cm^{-1}, they are not very sensitive to the molecular environment, the three methyl groups are indiscernible with a resolution of 2 cm^{-1}. Already we must remark that two out-of-plane modes have very low frequencies at 70 and 74 cm^{-1} in the range of the lattice modes. In the neutron scattering spectra there are two ranges which are very sensitive to isotopic substitution and are particularly intense, 45-85 cm^{-1} and 140-165 cm^{-1}, they were treated as giving signs of excitations of the methyl quantum rotors from their ground state to the first and second librational levels.

QUANTUM LEVELS OF THE METHYL ROTORS AND STRUCTURAL CONSEQUENCES

Recalling earlier results (4) (5) (6), about the tunnelling of the three methyl groups of the tribromomesitylene molecules, we have found three tunnelling excitations around 14, 25 and 49 µeV, this demonstrates that even if the molecules in the crystal have a C_{3h} symmetry the differences of environment of the methyl groups in the triclinic cell are sufficient to give different hindering potentials V_h for each methyl group. From these observations and with the assumption of the assignments of the table 3.a, for each methyl group we have calculated potentials of the shape $V_h = (V_3/2) \cos 3\Theta + (V_6/2) \cos(6\Theta - \Phi)$ and from these potentials we have found the characteristic wave functions, we give lower down in table 4 the potential used and the resulting expressions of the functions Ψ_{Ao} for the ground state, these functions allowed us to calculate D the density probability for the protons on a circle C of radius 1.02 Å. This circle corresponds to a constant length 1.08 Å for the three C-H bonds rotating around one C-C bond, their symmetry axis. The Figure 2 gives this density $D = \Psi_{Ao} \cdot \Psi_{Ao}^*$ for the less hindered methyl group labelled Me n°7 when the circle is developped along a straight line. This calculated curve D is compared with the experimental data obtained on Figure 1 by the intersection of the circle C with the "iso-protonic density" curves and represented in Figure 2 by small squares. The densities calculated for the ground state level of symmetry Eo are superimposible to those calculated for Ao.

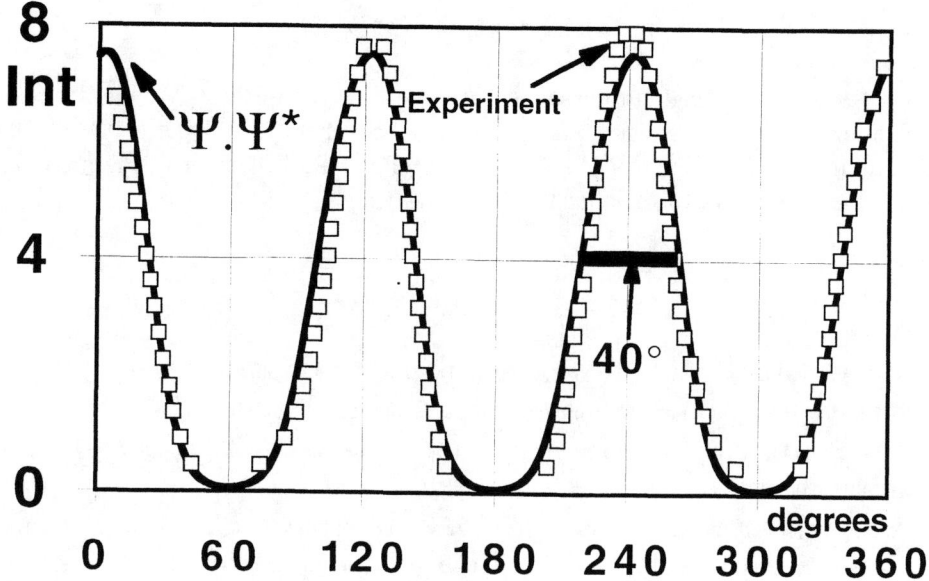

FIGURE 2. Protonic probability density for the methyl "7" of TBM at 14 K

The agreement calculation-experiment is very satisfactory, it is also good for the two other methyl groups. The calculated curve is only slightly enlarged when we take also in account the Debye-Waller factor due to the mean residual molecular motions at 14 K, we think that this result shows that the large half-height width of protonic probability density seen in the structural determination is the consequence of the quantum coherence of the protons in the methyl group.

TABLE 4. Hindering potentials and wave function of the ground state of symmetry A

Methyl 7
V_3=22.3meV V_6=8.7meV ϕ=186° $h\nu_t$=49µeV A_{01}=7.08meV E_{01}=6.66meV
Ψ_A= 0,849 +0.746cos3θ +0.034 sin3θ+0.012 cos6θ +0.020 sin6θ -0.027 cos9θ..
Methyl 8
V_3=32.0meV V_6=8.0meV ϕ=174° $h\nu_t$=14.0µeV A_{01}=9.24meV E_{01}=9.06meV
Ψ_A= 0,796 +0.848 cos3θ -0.026 sin3θ +0.112 cos6θ -0.018 sin6θ -0.014 cos9θ..
Methyl 9
V_3=24.8meV V_6=12.2meV ϕ=129° $h\nu_t$=25µeV A_{01}=8.80meV E_{01}=8.47meV
Ψ_A= 0,797 +0.777 cos3θ -0.292 sin3θ +0.052 cos6θ -0.196 sin6θ -0.032 cos9θ..

ACKNOWLEDGMENTS

We acknowledge Professor R. Carrié and Doctor M. Vaultier for the hospitality in their chemistry laboratory and hepful advises, Doctors C. Carlile, J.Tomkinson, M. Adams and A. Stewart for help at RAL, Doctor B. Hennion for help at LLB and Doctor M.Johnson for help at ILL.

REFERENCES

1. M. Mani, W. Paulus, A. Cousson and J. Meinnel, *J. Chem. Phys.*, submitted 1999
2. F. Boudjada, J. Meinnel and M. Sanquer, *Spectrochemica Acta*, submitted 1999
3. J. Meinnel, M. Mani, F. Tonnard, M. Nusimovici and M. Sanquer, *C.R. Acad. Sci. Paris* **317** II, 885- 890 (1993)
4. J. Meinnel, W. Häusler, M. Mani, M. Tazi, M. Nusimovici, M. Sanquer, B. Wyncke, A. Heidemann, C.J. Carlile, J. Tomkinson and B. Hennion, *Physica B* **181**, 711- 713 (1992)
5. J. Meinnel, M. Mani, M. Nusimovici, C.J. Carlile, B. Hennion, R. Carrié, B. Wyncke, M. Sanquer and F. Tonnard, *Physica B* **202**, 293- 301 (1994)
6. J. Meinnel, C.J. Carlile, K.S. Knight and J. Godard, *Physica B* **226**, 238- 240 (1996)
7. M. Prager and A. Heidemann, *Chem. Reviews*, **97-82**, 2933- 2966 (1997)
8. C.J. Carlile, *Int. J. Modern Phys. B* **7**, 3113- 3151 (1993)

Neutrons & Numerical Methods
Institut Laue-Langevin
Grenoble, France
9 - 12 December, 1998

Programme

Registration Desk opening hours : Wednesday 9H to 12H and 13H to 18H
Thursday and Friday 8H30 to 12H30 and 13H30 to 17H
Saturday 8H30 to 12H30

Wednesday, 9 December 1998

12h00   *buffet / ILL19/20*

All talks will be in the Chadwick Amphitheatre, ILL4

14h00		Welcome
		Session I Chairperson **H.P. Trommsdorff**
14h15	M. Payne	"Structure and dynamics from first principles"
15h00	L.L. Daemen	"Calculation of instrument resolution functions......"
15h20	R. McGreevy	"RMCPOW : a new RMC method for modelling..."
15h40	S.R. Elliott	"Theoretical determination of the Ioffe-Regel limit."

16h00   *coffee break / ILL4 Entrance hall*

		Session II Chairperson **J. Tomkinson**
16h30	A. Navarro	"Density functional theory and Ab Initio methods."
17h10	V. Khavryutchenko	"Quantum chemistry toolset for simulation of........"
17h30	A.J. Ramirez	"Molecular dynamics simulations of inelastic........."
17h50	J. Horbach	"High frequency dynamics of amorphous silica......"

Thursday, 10 December 1998

		Session III Chairperson **H. Schober**
9h00	K. Parlinski	"Calculation of phonon dispersion curves..............."
9h40	R. Fouret	"Phonons of chalcopyrites"
10h00	J. Li	"Simulations of inelastic neutron scattering data....."

10h20   *coffee break / ILL 4 Entrance hall*

		Session IV Chairperson **A. Deriu**
10h50	G. Kneller	"Analysis of low frequency motions in proteins ...."
11h30	N. Leulliot	"Search for a reliable nucleic acid force field using..."
11h50	A. Paciaroni	"Molecular dynamics simulation of inelastic.........."

12h30   *lunch / H2 Cantine Supérieure*

		Session V Chairperson **M. Payne**
14h00	S. French	"Molecular modelling of organic superconducting..."
14h40	O. Petrenko	"Geometric frustration in gadolinium gallium.........."
15h20	G. Prigent	"Local order and metal-non-metal transition in ......"
15h40	J. Colmenero	"On the origin of the distribution of potential........."

16h00   *coffee break / ILL4 Entrance hall*

16h30   Posters Session A / *ILL 19/20*

Friday, 11 December 1998

| | | Session VI | Chairperson F. Leclercq |

9h00	J. Gale	"Structure and properties of ionic materials from ..."
9h40	M. Ghomi	"On the obtention of a harmonic force field for........"
10h00	S.N. Taraskin	"The correction function for the derivation of the..."

10h20   coffee break / ILL 4 Entrance hall

10h40   Posters Session B / *ILL 19/20*

12h30   lunch / H2 Cantine Supérieure

| | | Session VII | Chairperson W. F. Kuhs |

14h00	G. Engel	"Using molecular modeling in crystal structure......."
14h40	N.D. Morelon	"Dynamics of a channel-type inclusion compound."
15h20	B. Chazallon	"Molecular dynamics modelling and neutron ...."
15h40	V. Coulet	"Neutron scattering study of S-Te liquid alloys......"

16h00   coffee break / ILL 4 Entrance hall

| | | Session VIII | Chairperson G. Kneller |

16h30	S.J. Grabowski	"Simulations of hydrogen bonds in crystals and ....."
16h50	M. Dibari	"Molecular dynamics simulations and quasi-elastic"
17h10	M. Chahid	"D_2O in zeolites type Faujasite"
17h40	D.H. Powell	"Neutron scattering studies of the structure and....."

20h00   dinner, Château de Sassenage

Saturday, 12 December 1998

| | | Session IX | Chairperson M. Bée |

9h00	M. Guest	"Algorithms, developments and applications in......"
9h40	T. Yildirim	"Combined neutron scattering and first principles.."
10h00	M. Neumann	"Probing potential energy surfaces by methyl ...."

10h20   coffee break / ILL 4 Entrance hall

| | | Session X | Chairperson R. McGreevy |

10h50	L. Borjesson	"The structure of many component glasses using..."
11h30	J. Swenson	"A new approach to investigation migration..........."
11h50	S. Kugler	"Molecular dynamics simulation of the growth of.."
12h10	Conclusions	

12h30   buffet / ILL 4 Entrance hall

LIST OF PARTICIPANTS

Surname : ALBINATI
First Name : Alberto
Address : INST.PHARM. CHEMISTRY
UNIVERSITA DI MILANO
42 VIALE ABRUZZI
I-20131 MILANO

Tel : +39 02 29523911
Fax : +39 02 29514197
E-Mail : albinati@unimi.it

Surname : ARAI
First Name : Takashi
Address : NAT DEFENSE ACADEMY
YOKOSUKA 239
JAPAN

Tel : +81 468 41 3810
Fax : +81 468 44 5902
E-Mail : arai@cc.nda.ac.jp

Surname : AUBERT
First Name : Brigitte
Address : INSTITUT MAX VON LAUE - PAUL L
BP 156
F-38042 GRENOBLE CEDEX 9

Tel : +33 4 76 20 70 08
Fax : +33 4 76 48 39 06
E-Mail : aubert@ill.fr

Surname : BEE
First Name : Marc
Address : LABORATOIRE DE SPECTRO.
PHYSIQUE
UNIVERSITE JOSEPH FOURIER
B.P. 87
F-38402 ST.MARTIN D'HERES

Tel : +33 4 76 51 47 60
Fax : +33 4 76 51 45 44
E-Mail : marc.bee@ujf-grenoble.fr

Surname : BENDERSKII
First Name : V.A.
Address : INSTITUTE CHEM PHYSICS
RUSSIAN ACADEMY OF SCIENCE
142432 MOSCOW REGION
CHERNOGOLOVKA
RUSSIA

Tel :
Fax :
E-Mail : vbenders@spectro.ujf-grenoble.fr

Surname : BIZZARRI
First Name : Anna Rita
Address : DIP SCIENZE AMBIENTALI
SEZIONE CHIMICA E FISICA
UNIVERSITA DELLA TUSCIA
I-01100 VITERBO

Tel : +39 761357136
Fax : +39 761357179
E-Mail : annarita.bizzarri@pg.infn.it

Surname : BORJESSON
First Name : Lars
Address : DEPT OF APPLIED PHYSICS
CHALMERS UNIV TECHNOLOGY
S-412 96 GOTHENBURG
SWEDEN

Tel : +46 31 772 33 07
Fax : +46 31 772 20 90
E-Mail : borje@fy.chalmers.se

Surname : BOUDJADA
First Name : Fahima
Address : UNIV RENNES
BAT. 11A - GMCM
CAMPUS DE BEAULIEU
AVENUE GAL LECLERC
F-35042 RENNES CEDEX

Tel : +33 02 99 28 60 58
Fax : +33 02 99 28 67 17
E-Mail : boudjada@univ-rennes1.fr

Surname: BUTTNER First Name: Herma G. Address: INSTITUT MAX VON LAUE - PAUL L BP 156 F-38042 GRENOBLE CEDEX 9 Tel: +33 4 76 20 7179 Fax: +33 4 76 4839 06 E-Mail: buttner@ill.fr	Surname: CANNISTRARO First Name: Salvatore Address: DIPARTIMENTO DI SCIENZA SEZIONE CHIMICA E FISICA UNIVERSITA DELLA TUSCIA I-01100 VITERBO Tel: +39 075 585 3032 Fax: +39 075 44 666 E-Mail: cannistraro@pg.infn.it
Surname: CHAHID First Name: Abdellah Address: STUDSVIK NEUTRON RESEARCH LABORATORY S-611 82 NYKOEPING SWEDEN Tel: +46 155 221840 Fax: +46 155 263001 E-Mail: chahid@studsvik.uu.se	Surname: CHAZALLON First Name: Bertrand Address: MKI UNIVERSITAET GOETTINGEN GOLDSCHMIDTSTR. 1 D-37077 GOETTINGEN Tel: +49 551 39 38 92 Fax: +49 551 39 95 21 E-Mail: bertrand@silly.uni-mki.dwdg.de
Surname: COLMENERO First Name: Juan Address: UNIV PAIS VASCO DEPT FISICA MAT APDO. 1072 E-20080 SAN SEBASTIAN Tel: +34 943 44 82 05 Fax: +34 943 21 22 36 E-Mail: wapcolej@sc.ehu.es CHAIRPERSON	Surname: COMBET First Name: Jerome Address: INSTITUT MAX VON LAUE - PAUL LANGEVIN B.P. 156 F-38042 GRENOBLE CEDEX 9 Tel: +33 4 76 20 75 68 Fax: +33 4 76 48 39 06 E-Mail: combet@ill.fr
Surname: COULET First Name: Marie Vanessa Address: CNRS/CTM 26 RUE DU 141E RIA F-13003 MARSEILLE Tel: +33 4 912 82078 Fax: +33 4 915 03829 E-Mail: vcoulet@ctm.cnrs-mrs.fr	Surname: COUSSON First Name: Alain Address: LABORATOIRE LEON BRILLOUIN CEN SACLAY F-91191 GIF-SUR-YVETTE CEDEX Tel: +33 1 69 08 64 31 Fax: +33 1 69 08 82 61 E-Mail: cousson@bali.saclay.cea.fr

Surname : CRACIUN First Name : Steliana Address : CNRS/LSP-SPECTRO PHYS UNIV GRENOBLE 1 BP 87 - BAT E45 140 RUE DE LA PHYSIQUE 38402 ST MARTIN D'HERES Tel : +33 4 76 51 48 73 Fax : +33 4 76 51 45 44 E-Mail : craciun@spectro.ajf-grenoble.fr	Surname : CZIHAK First Name : Christoph Address : INSTITUT MAX VON LAUE - PAUL LANGEVIN B.P. 156 F-38042 GRENOBLE CEDEX 9 Tel : +33 4 76 20 73 99 Fax : +33 4 76 48 39 06 E-Mail : czihak@ill.fr
Surname : DAEMEN First Name : Luke L. Address : LOS ALAMOS NAT LAB LOS ALAMOS NEW MEXICO 87545 USA Tel : +505 667 9695 Fax : +505 665 2676 E-Mail : lld@lanl.gov	Surname : DAMAY First Name : Pierre Address : LASIR - CNRS HEI 13 RUE DE TOUL F-59046 LILLE CEDEX Tel : +33 3 28 38 48 58 Fax : +33 3 28 38 48 04 E-Mail : pierre.damay@hei.fr
Surname : DAWIDOWSKI First Name : Javier Address : CENT ATOM BARILOCHE 8400 SAN CARLOS DE BARILOCHE RIO NEGRO ARGENTINA Tel : +54 944 45216 Fax : +54 944 45299 E-Mail : javier@cab.cnea.edu.ar	Surname : DELAPLANE First Name : Robert Address : THE STUDSVIK NEUTRON RES. LAB UNIV UPPSALA S-611 82 NYKOEPING SWEDEN Tel : +46 155 22 1843 Fax : +46 155 26 3001 E-Mail : rgd@studsvik.uu.se
Surname : DERIU First Name : Antonio Address : DIP DI FISICA UNIVERSITA DI PARMA VIALE DELLE SCIENZE I-43100 PARMA Tel : +39 0521 905267 Fax : +39 0521 905223 E-Mail : Antonio.Deriu@fis.unipr.it	Surname : DIANOUX First Name : Albert Jose Address : INSTITUT MAX VON LAUE- PAUL LANGEVIN BP 156 F-38042 GRENOBLE CEDEX 9 Tel : +33 4 76 20 72 06 Fax : +33 4 76 48 39 06 E-Mail : dianoux@ill.fr

Surname : DIBARI First Name : MARIA TERESA Address : UNIV PARMA PARCO AREA DELLE SCIENZE, 7A PARMA I-43100 ITALY Tel : +39 521 905 561 Fax : +39 521 905 223 E-Mail : mariateresa.dibari@fis.unipr.it	Surname : ELLIOTT First Name : Stephen R. Address : DEPT OF CHEMISTRY UNIV CAMBRIDGE LENSFIELD ROAD CAMBRIDGE CB2 IEW GRANDE-BRETAGNE Tel : +44 1223 336525 Fax : +44 1223 336362 E-Mail : srel@cus.cam.ac.uk
Surname : ENGBERG First Name : Dennis Address : ISIS RUTHERFORD APPLETON LAB. CHILTON, DIDCOT OXON OX11 0QX ENGLAND Tel : +44 1235 445797 Fax : +44 1235 445720 E-Mail : d.engberg@rl.ac.uk	Surname : ENGEL First Name : Gerhard Address : MOLECULAR SIMULATIONS LTD 230/250 THE QUORUM BARNWELL ROAD CAMBRIDGE CB5 8RE U.K. Tel : +44 1223 507524 Fax : +44 1223 413301 E-Mail : gengel@msicam.co.uk
Surname : FOURET First Name : Rene Address : UNIVERSITE DE LILLE I BATIMENT P5 F-59655 VILLENEUVE D'ASCQ CEDE Tel : +33 3 20 43 47 73 Fax : +33 3 20 43 40 84 E-Mail : rene.fouret@univ-lille1.fr	Surname : FRENCH First Name : Sam Address : ROYAL INSTITUTION 21 ALBEMARLE STREET LONDON W1X 4BS GRANDE-BRETAGNE Tel : +44 171 409 2992 Fax : +44 171 670 2920 E-Mail : sam@ri.ac.uk
Surname : FRICK First Name : Bernhard Address : INSTITUT MAX VON LAUE - PAUL LANGEVIN BP 156 F-38042 GRENOBLE CEDEX 9 Tel : +33 4 76 20 73 22 Fax : +33 4 76 48 39 06 E-Mail : frick@ill.fr ILL SPECIALIST	Surname : GALE First Name : Julian Address : DEPT OF CHEMISTRY IMPERIAL COLLEGE OF SCIENCE TECHNOLOGY & MEDICINE SOUTH KENSINGTON SW7 2AY LONDON UK Tel : +44 171 594 5757 Fax : +44 171 594 5804 E-Mail : j.gale@ic.ac.uk

Surname : GAY-DUCHOSAL First Name : Michael Address : ICMA UNIVERSITE DE LAUSANNE-BCH CH-1015 LAUSANNE Tel : +41 21 692 3862 Fax : +41 21 692 3855 E-Mail : michael.gay-duchosal@icma.unil.ch	Surname : GHOMI First Name : Mahmoud Address : LPBC-LAB PHYS BIOMOL CELL UNIVERSITE P. & M. CURIE CASE COURRIER 138 4 PLACE JUSSIEU 75252 PARIS CEDEX 5 Tel : +33 1 44 27 7555 Fax : +33 1 44 27 7560 E-Mail : ghomi@lpbc.jussieu.fr
Surname : GHOSH First Name : Ronen Address : I.L.L. BP 156 38042 GRENOBLE CEDEX 9 FRANCE Tel : +33 4 76 20 71 78 Fax : +33 4 76 48 39 06 E-Mail : ron@ill.fr	Surname : GRABOWSKI First Name : Slawomir J. Address : INSTITUTE OF CHEMISTRY UNIVERSITY IN BIALYSTOK 15-443 BIALYSTOK AL. PILSUDSKIEGO 11 POLAND Tel : 048 457 121 Fax : E-Mail : slagra@noc.uwb.edu.pl
Surname : HANHAM First Name : Mathew Address : DEPT OF PHYSICS UNIV OF WARWICK COVENTRY CV4 7AL GRANDE-BRETAGNE Tel : +44 1203 522235 Fax : E-Mail : hanham@trebor.phys.warwick.ac.uk	Surname : HATHARASINGHE First Name : Mallika Address : DEPT PHYS & ASTRON UNIV COLLEGE LONDON GOWER STREET LONDON WC1E 6BT GRANDE BRETAGNE Tel : +44 171 419 3413 Fax : +44 171 391 1360 E-Mail : h.hatharasinghe@ucl.ac.uk
Surname : HAYAMA First Name : Shusaku Address : DEPT PHYS & ASTRON UNIV COLLEGE LONDON GOWER STREET LONDON WC1E 7BT GRANDE BRETAGNE Tel : +44 171 7147 (X3505) Fax : +44 171 380 7145 E-Mail : ucap@ucl.ac.uk	Surname : HOLDERNA-NATKANIEC First Name : Krystyna Address : INST PHYS UNIVERSITY MICKIEWICZ UMULTOWSKA 85 61-614 POZNAN POLAND Tel : +48 061 821 7011 Fax : +48 061 821 7981 E-Mail : natkanie@main.amu.edu.pl

Surname :	HORBACH
First Name :	Juergen
Address :	JOHANNES GUTENBERG UNIVERSITY MAINZ STAUDINGER WEG 7 D-55099 MAINZ
Tel :	+49 6131 393643
Fax :	+49 6131 395441
E-Mail :	horbach@knuddel.physik.uni-mainz.de

Surname :	JOHNSON
First Name :	Mark Robert
Address :	INSTITUT MAX VON LAUE-PAUL LANGEVIN B.P. 156 F-38042 GRENOBLE CEDEX 9
Tel :	+33 4 76 20 71 39
Fax :	+33 4 76 48 39 06
E-Mail :	johnson@ill.fr

Surname :	KEARLEY
First Name :	Gordon
Address :	I.L.L. BP 156 F-38042 GRENOBLE CEDEX 9
Tel :	+33 4 76 20 71 33
Fax :	+33 4 76 48 39 06
E-Mail :	kearley@ill.fr

Surname :	KHAVRYUTCHENKO
First Name :	Vladimir
Address :	NAT ACADEMY OF SCIENCE COMPUTATION CHEM GROUP INST SURFACE CHEMISTRY 252650 KIEV, UKRAINE
Tel :	+44 563 4274
Fax :	+44 228 0060
E-Mail :	vkhavr@compchem.kiev.ua

Surname :	KLAPPROTH
First Name :	Alice
Address :	MIN KRIST INSTITUT UNIV GOETTINGEN GOLDSCHMIDTSTRASSE 1 D-37077 GOETTINGEN
Tel :	+49 551 393892
Fax :	+49 551 399521
E-Mail :	alice@silly.uni-mki.gwdg.de

Surname :	KNELLER
First Name :	Gerald
Address :	CENTRE BIOPHYS MOLECULAIRE RUE CHARLES SADRON F-45071 ORLEANS CEDEX 2
Tel :	+33 2 38257842
Fax :	+33 2 38631517
E-Mail :	kneller@cnrs-orleans.fr

Surname :	KUGLER
First Name :	Sandor
Address :	INSTITUTE OF PHYSICS TECHNICAL UNIVERSITY OF BUDAPEST H-1521 BUDAPEST HUNGARY
Tel :	+36 1 463 2146
Fax :	+36 1 463 3567
E-Mail :	kugler@phy.bme.hu

Surname :	KUHS
First Name :	Werner F.
Address :	MINERALOGISCH-KRISTALLO. INSTITUT UNIVERSITAET GOETTINGEN GOLDSCHMIDTSTR 1 D-37077 GOETTINGEN
Tel :	+49 551 39 38 91
Fax :	+49 551 39 95 21
E-Mail :	kuhs@silly.uni-mki.gwdg.de CHAIRPERSON

Surname : LECHNER First Name : R.E. Address : HAHN-MEITNER-INSTITUT ABT. B.E.N.S.C. GLIENICKERSTRASSE 100 D-14109 BERLIN Tel : +49 30 8062 2780 Fax : +49 30 80622181 E-Mail : lechner@hmi.de	Surname : LECLERCQ First Name : Francoise Address : LABORATOIRE DE CHIMIE PHYSIQUE URP 2631 DU CNRS - HEI 13, RUE DE TOUL F-59046 LILLE CEDEX Tel : +33 3 28 38 48 13 Fax : +33 3 28 38 48 04 E-Mail : francoise.leclercq@hei.fr
Surname : LEULLIOT First Name : Nicolas Address : CNRS/LPBC-LAB PHYS BIOMOL CEL UNIVERSITE P. & M. CURIE CASE COURRIER 138 4 PLACE JUSSIEU 75252 PARIS CEDEX 5 Tel : +33 1 44 27 7557 Fax : +33 1 44 27 7560 E-Mail : leulliot@lpbc.jussieu.fr	Surname : LI First Name : Jichen Address : UNIVERSITY OF MANCHESTER DEPARTMENT OF PHYSICS PO BOX 88 MANCHESTER M60 1QD GRANDE-BRETAGNE Tel : +44 161 200 3933 Fax : +44 161 200 3941 E-Mail : j.c.li@umist.ac.uk
Surname : MADSEN First Name : Georg Address : DEPT CHEM UNIVERSITY OF AARHUS LANGELANDSGADE 140 DK-8000 AARHUS C Tel : +45 89 42 38 93 Fax : +45 86 19 61 99 E-Mail : georgm@kemi.aau.dk	Surname : MCGREEVY First Name : Robert Address : STUDSVIK NEUTRON RES. LAB. UPPSALA UNIVERSITY S-611 82 NYKOEPING SWEDEN Tel : +46 155 221831 Fax : +46 155 263001 E-Mail : mcgreevy@studsvik.uu.se
Surname : MEDINI First Name : Duccio Address : DIPART FISICA UNIV PERUGIA VIA A. PASCOLI I-06100 PERUGIA Tel : +39 75 5853060 Fax : +39 75 44666 E-Mail : duccio.medini@pg.infn.it	Surname : MEINNEL First Name : Jean Address : UNIV RENNES I CAMPUS BEAULIEU BAT. 22, AV. GEN. LECLERC F-35042 RENNES CEDEX Tel : +33 2 99 28 60 58 Fax : +33 2 99 28 67 17 E-Mail : meinnel@univ-rennes1.fr

Surname : MIDDENDORF First Name : H.D. Address : CLARENDON LABORATORY OXFORD UNIVERSITY PARKS ROAD OXFORD OX1 3PU GRANDE BRETAGNE Tel : +44 1442 863 828 Fax : +44 1865 272400 E-Mail : hdm01@isise.rl.ac.uk	Surname : MITCHELL First Name : Philip C.H. Address : DEPT OF CHEMISTRY UNIV OF READING READING RG6 6D GREAT-BRITAIN Tel : +44 118 931 6344 Fax : +44 118 931 6331 E-Mail : scsmitch@reading.ac.uk
Surname : MORELON First Name : Nhan-Duc Address : INSTITUT MAX VON LAUE - PAUL LANGEVIN B.P. 156 F-38042 GRENOBLE CEDEX 9 Tel : +33 4 76 20 7567 Fax : +33 4 76 48 39 06 E-Mail : morelon@ill.fr	Surname : NATKANIEC First Name : Ireneusz Address : LABORATORY OF NEUTRON PHYSICS - JINR 141980 DUBNA RUSSIA Tel : +7 96 21 65448 Fax : E-Mail : inat@nf.jinr.ru
Surname : NAVARRO First Name : Amparo Address : DEPT QUIMICA FISICA FACULTAD DE CIENCIAS UNIVERSIDAD JAEN E-23071 JAEN Tel : +34 953 21 21 62 Fax : +34 953 21 25 26 E-Mail : anavarro@ujaen.es	Surname : NEUMANN First Name : Marcus Address : INSTITUT MAX VON LAUE - PAUL LANGEVIN B.P. 156 F-38042 GRENOBLE CEDEX 9 Tel : +33 4 76 20 71 39 Fax : +33 4 76 48 39 06 E-Mail : neumann@ill.fr
Surname : NICOLAI First Name : Beatrice Address : I.L.L BP 156 F-38042 GRENOBLE CEDEX 9 Tel : +33 4 76 20 72 32 Fax : +33 4 76 48 39 06 E-Mail : nicolai@ill.fr	Surname : PACIARONI First Name : Alessandro Address : DIPART FISICA UNIV PERUGIA VIA A. PASCOLI I-06100 PERUGIA Tel : +39 755853029 Fax : +39 7544666 E-Mail : paciaroni@ill.fr

Surname : PARLINSKI First Name : Krzysztof Address : UNIV PHYS NUCL PHYSICS UL. RADZIKOWSKIEGO 152 31342 CRACOW POLAND Tel : +48 12 637 0222 Fax : +48 12 637 5441 E-Mail : parlinski@vsb01.ifj.edu.pl	Surname : PAYNE First Name : Mike Address : CAVENDISH LAB MADINGLEY ROAD CAMBRIDGE CB3 0HE UNITED KINGDOM Tel : +44 1223 337381 Fax : +44 1223 337356 E-Mail : mcp1@phy.cam.ac.uk
Surname : PETRENKO First Name : Oleg Address : DEPT PHYSICS UNIV WARWICK COVENTRY CV4 7AL GRANDE-BRETAGNE Tel : +44 1203 52 34 14 Fax : +44 1203 69 20 16 E-Mail : phsby@csv.warwick.ac.uk	Surname : PHILLIPS First Name : Jon R. Address : DEPT OF PHYSICS UNIV NOTTINGHAM UNIVERSITY PARK NOTTINGHAM NG7 2RD GRANDE-BRETAGNE Tel : +44 115 951 5151 Fax : +44 115 951 5180 E-Mail : ppxjrp@nottingham.ac.uk
Surname : PITTELOUD First Name : Cedric Address : ICMA UNIVERSITE DE LAUSANNE CH-1015 LAUSANNE Tel : +41 21 692 3929 Fax : +41 21 692 3935 E-Mail : cpittelo@icma.unil.ch	Surname : PLAZANET First Name : Marie Address : CNRS/LSP-SPECTRO PHYS UNIV GRENOBLE 1 BP 87 - BAT E45 140 RUE DE LA PHYSIQUE 38402 ST MARTIN D'HERES Tel : +33 4 76 51 48 73 Fax : +33 4 76 51 45 44 E-Mail : plazanet@spectro.ujf-grenoble.fr
Surname : POWELL First Name : D. Hugh Address : ICMA UNIVERSITE DE LAUSANNE CH-1015 LAUSANNE Tel : +41 21 692 39 18 Fax : +41 21 692 39 25 E-Mail : hpowell@icma.unil.ch	Surname : PRIGENT First Name : Gilles Address : LAB LEON BRILLOUIN CEA SACLAY BATIMENT 563 91191 GIF-SUR-YVETTE CEDEX Tel : +33 1 69 08 49 54 Fax : +33 1 69 08 82 61 E-Mail : prigent@bali.saclay.cea.fr

Surname :	RAMIREZ-CUESTA
First Name :	Anibal Javier
Address :	UNIVERSITY OF READING
	DEPT. OF PHYSICS
	WHITEKNIGHTS
	PO BOX 224, READING RG6 6AD
	GRANDE-BRETAGNE
Tel :	+44 118 87 5123
Fax :	+44 118 931 6331
E-Mail :	a.j.ramirez-cuesta@reading.ac.uk

Surname :	REAT
First Name :	Valerie
Address :	ILL
	INSTITUT MAX VON LAUE -
	PAUL LANGEVIN
	B.P. 156
	F-38042 GRENOBLE CEDEX 9
Tel :	
Fax :	
E-Mail :	reat@ill.fr

Surname :	RUSSINA
First Name :	Margarita
Address :	LOS ALAMOS NAT LAB
	LOS ALAMOS
	NEW MEXICO 87545
	USA
Tel :	+1 505 667 8841
Fax :	+1 505 665 2676
E-Mail :	russina@popler.lansce.lanl.gov

Surname :	RYCROFT
First Name :	Stuart
Address :	NFL STUDSVIK
	UPPSALA UNIVERSITY
	S-611 82 NYKOEPING
	SWEDEN
Tel :	+46 155 221871
Fax :	
E-Mail :	stuart@studsvik.uu.se

Surname :	SCHEIDLER
First Name :	Peter
Address :	UNIVERSITAET MAINZ
	JOHANNES GUTENBERG
	INST. F. PHYSIK
	WA 331
	D-55099 MAINZ
Tel :	+49 6131 395 158
Fax :	+49 6131 395 441
E-Mail :	peter.scheidler@uni-mainz.de

Surname :	SCHOBER
First Name :	Helmut
Address :	INSTITUT MAX VON LAUE
	PAUL LANGEVIN
	BP 156
	38042 GRENOBLE CEDEX 9
Tel :	+33 4 76 20 73 96
Fax :	+33 4 76 48 39 06
E-Mail :	schober@ill.fr

Surname :	SWENSON
First Name :	Jan
Address :	DEPT APPL PHYSICS
	CHALMERS UNIV OF TECHNOLOGY
	S-412 96 GOETEBORG
Tel :	+46 31 772 5680
Fax :	+46 31 772 2090
E-Mail :	f5xjs@fy.chalmers.se

Surname :	TARASKIN
First Name :	Sergei
Address :	DEPT CHEM
	UNIV CAMBRIDGE
	LENSFIELD ROAD
	CAMBRIDGE CB2 IEW
	GRANDE-BRETAGNE
Tel :	+44 1223 336532
Fax :	+44 1223 336362
E-Mail :	snt1000@cus.cam.ac.uk

Surname : TAREK First Name : Mounir Address : NIST MARYLAND A 141 PHYSICS BLDG GAITHERSBURG MARYLAND MD 20899-0001 U.S.A. Tel : +301 975 3959 Fax : +301 921 9847 E-Mail : tarek@rrdjazz.nist.gov	Surname : TOMKINSON First Name : John Address : ISIS CCLRC RAL CHILTON DIDCOT OXON OX11 0QX GRANDE-BRETAGNE Tel : +44 1235 44 6686 Fax : +44 1235 44 5383 E-Mail : jt@isise.rl.ac.uk
Surname : TROMMSDORFF First Name : Hans-Peter Address : LAB SPECTRO PHYSIQUE UNIV JOSEPH FOURIER 38402 ST MARTIN D'HERES CEDEX Tel : +33 4 76 51 47 81 Fax : +33 4 76 51 45 44 E-Mail : hans-peter.trommsdorff@ujf-grenoble.fr	Surname : YILDIRIM First Name : Taner Address : NIST CENTER FOR NEUTRON RES BLDG. 235, RM E112 GAITHERSBURG MARYLAND MD 20899 U.S.A. Tel : +301 975 6228 Fax : +309 921 9847 E-Mail : taner@nist.gov
Surname : ZETTERSTROEM First Name : Per Address : THE STUDSVIK NEUTRON RES. LAB S-61182 NIKOEPING SWEDEN Tel : +46 155 221837 Fax : +46 155 263001 E-Mail : perz@studsvik.uu.se	

Author Index

A

Alegría, A., 201
Alvarez, F., 201

B

Baron, M.-H., 183
Bellissent, R., 83
Bichara, C., 83
Binder, K., 131, 136
Bizzarri, A. R., 142
Blostein, J. J., 37
Börjesson, L., 57
Boudjada, F., 217
Büttner, H. G., 102

C

Cannistraro, S., 142
Catlow, C. R. A., 96
Chazallon, B., 70, 74
Colmenero, J., 201
Cousson, A., 217

D

Daemen, L. L., 41
Davies, G. R., 201
Dawidowski, J., 37
Derollez, P., 127

E

Elliott, S. R., 160

F

Fernández-Gómez, M., 172
Fouret, R., 127
French, S. A., 96

G

Gale, J. D., 28
Gaspard, J.-P., 83
Gay-Duchosal, M., 78
Ghomi, M., 179, 183
Gonzalez, J., 127
Grabowski, S. J., 112
Grajcar, L., 183
Granada, J. R., 37
Guest, M. F., 9

H

Haar, A., 160
Hanham, M. L., 167
Hennion, B., 127
Hinsen, K., 147
Hjelm, R. P., 41
Holderna-Natkaniec, K., 187, 191
Horbach, J., 131, 136

J

Jobic, H., 179, 183
Johnson, M. R., 212

K

Kalus, J., 191
Kearley, G. J., 102, 172, 206
Khavryutchenko, V. D., 187, 191
Klapproth, A., 70, 74
Kneller, G. R., 147
Kob, W., 131, 136
Koháry, K., 64
Kugler, S., 64
Kuhs, W. F., 70, 74

L

Laamyem, A., 127
László, I., 64
Leulliot, N., 179, 183

Li, J., 155
López-González, J. J., 172

M

Madsen, G. K. H., 107
Mani, M., 217
McGreevy, R. L., 19, 57
McK. Paul, D., 90
Meinnel, J., 217
Mellergård, A., 19
Mezei, F., 47
Mitchell, P. C. H., 195

N

Natkaniec, I., 187, 191
Navarro, A., 172
Neumann, M. A., 212
Nicholson, T. M., 201
Nicolai, B., 206

O

Oeffner, R. D., 160

P

Paciaroni, A., 142
Parker, S. F., 195
Parlinski, K., 121
Partal, F., 172
Paulus, W., 217
Payne, M. C., 3
Petrenko, O. A., 90
Pettifer, R. F., 167
Pitteloud, C., 78
Plazanet, M., 212

Powell, D. H., 78
Prager, M., 102
Prandl, W., 102
Prigent, G., 83

R

Ramirez-Cuesta, A. J., 195
Rodger, P. M., 195
Russina, M., 47

S

Sanquer, M., 217
Scheidler, P., 131
Schiebel, P., 102
Seeger, P. A., 41
Swenson, J., 57

T

Taraskin, S. N., 160
Thelliez, T. G., 41
Tomkinson, J., 155, 172

U

Ulicny, J., 183

V

van Dam, H. J. J., 9

W

Wilkinson, A. P., 195